Softcover ISBN 978-1-957077-99-4
Hardcover ISBN 978-1-957077-00-0
Cover image: Shutterstock 58727S103

HARVEST OF HEALING, LLC

IZAИA 61®

Publishing assistance by BookCrafters, Parker, Colorado.
www.bookcrafters.net

Owner's Manual of the Universe

The Owner's Manual written by the Universe for the human life must be fully recovered. Lost through the sands of time resulting in disease, decay and death, the ancient language spoken by the Universe is gradually experiencing its long-awaited resurrection. The laws that govern mankind, those that orchestrate health and longevity, are being brought forth by the Master. Now is the time to learn and apply the language all of mankind desperately needs to know.

Izauh 61®

In an Attempt to Save the World

THE POWERFUL INFLUENCE OF CLOTHING:
COLOR, FABRIC & STYLE

HARVEST OF HEALING, LLC

Izauh 61®

*Too much knowledge causes people to miss out
on the conversation with Wisdom*

INDEX OF CONTENTS

NUTS & BOLTS

Terms used throughout this book reference an aspect of a similar or common term. The terms I use are to highlight that somehow, to some degree, there is an aspect, measurement of change or composition that is different from the familiar term and its definition. Beginning with the publication of From AntiChrist to I AM, unique terms have been used and referenced throughout the various publications to reflect this "similar, yet not" product or situation. Keep in mind, mankind was created to assist planet earth in its ebbs, flows and cyclical life. To keep that life cycle moving in a healthy direction, mankind must learn how to keep their physical body healthy, not only for their own benefit but for the benefit of the environment in which all living creatures exist. I call this healthy, photosynthesis producing human body a Spiritual Royal, one who does work in an unseen manner and in the process has a highly respected value; the person whose body is self-sustaining enough that it can assist others and the environment in overcoming a setback. For planet earth to keep on living and keep on spinning, a collective group of individuals must learn how to work and live by the laws of the Universe that results in an epic dance with the Spirit.

The terms dirty or muddy water are used to describe an aspect of the internal fluid(s), likely originating in the plasma then into the spine and moving through the channels of process to the cells and captured in the lymph, that has become negatively influenced, contaminated. Many times, when fluid in the body is dirty or contaminated, it is flying under the radar of standardized medical testing and brews in the body until an infection or cancer manifests. It is as though the fluid is at a "pre-infection" status and remains undetected. When certain foods are eaten, certain clothing is worn and/or certain colors dominate the environment you spend most of your time in, the fluid in the body will become impacted. Outside of the standard White Blood Cell count (WBC), that impact is not measurable and lies unnoticed until a serious health condition erupts. From my own personal experience, the calculations for "out of range" blood tests needs adjusted if we are to capture those simmering health issues that lead to death. Many blood related infections or diseases become out of control before medical providers give notice with the current standard reference range in place. Cancers can be treated by standard and non-standard means and then return months or years later in part due to this contaminated, dirty water issue. Even more concerning is newborns can have a level of inherited plasma contamination. The story in Numbers 5:11-31 is about stepping outside of Universal Law restrictions when it comes to food (barley is referenced in the story) that contains protein. Protein is a foundation for cells and when an incorrect protein is consumed and becomes the structure for new cells, the cells have a fragment of contamination or disability. When this happens, the Heavenly Gases (described below) become disrupted or chaotic, an alchemy interruption is birthed that eventually leads to emotional eruptions and/or physical decline.

Chemical elements and Heavenly Gases are terms used. When chemical elements are referenced, it encompasses the whole of all chemical elements. When the term Heavenly Gases is used it is a reference to the collection of gases that are absorbed and used by the physical body. Not knowing what that specific cluster of gases would be for the body and not being certain that those gases have been measured when inside the body, I call them Heavenly Gases. Therefore, Heavenly Gases is the term used until in the future when appropriate testing to reveal what gases are involved and in what measure is had.

Star Dust is, in part, the electrical charge produced by Heavenly Gases and cosmic activity that takes up residence in the plasma and eventually the cells.

INTRODUCTION

<u>The Flood</u>

As I gather and review the collection of notes marking my experiences through the years and sit down to write another segment from the wisdom gained, I notice there is a common theme of fluid (water) that rings throughout the various topics I have previously shared and am about to share. As I begin piecing the numerous points of information together, I wonder if we, collectively as human beings, have created situations where the fluid responsible for the natural cleansing process of the body has become contaminated. If the internal fluid, that makes up the spinal fluid and eventually cells, is full of filth the brain becomes a target for unwanted debris. Is this fluid <u>meant to be</u> a part of a healthy functioning immune system that has become so contaminated the body is thrown off course when it comes to fighting off invaders? The term Living Water is used in Scripture in the Holy Bible. It appears this Living Water is meant to be the fluid that is present in a human body although the fluid most humans currently have is far less than "Living." With Living Water, the human body will "thirst no more" by producing its own hydration and ability to eliminate harmful invaders. The majority of the population within the United States appear to be dying of thirst simply by the size of the water bottles they carry. When the plasma and corresponding fluids are clean and in proper functioning order (aka Living), the body will be and remain healthy.

John 7:37-38: On the last and most important day of the festival, Jesus stood up and cried out, "If anyone is thirsty, he should come to Me and drink! The one who believes in Me, as the Scripture has said, will have streams of living water flow from deep within him." (HCS)

This malfunctioning internal fluid system makes me think of a washing machine with a clog of dirty water in the drainage hose. Weekly our body should move through a cleansing process similar to how one would gather and sort dirty laundry to start a washing machine cycle for their clothing. If you are familiar with the True Sabbath Meditation shared in the Living by the Light of the Moon published 2024, you have been introduced to the cleansing and recharge process that takes place in the physical body from Friday evening through Monday early morning, the time to gather the dirty laundry and place it in the washing machine. On Saturday morning we start the washing machine and discover the machine is challenged at eliminating the dirty water from the load it was given. The drainage hose is unable to spill out the dirty water in the proper manner and therefore the dirty water recycles for each load or, seeps into areas that are not designed to handle or remove the water. Now we have clothes that are not clean and/or a flooding problem. A flood consisting of dirty water no less, that can often take up residence inside the cells. We all know that flooding causes various challenging messes and will eventually result in mold or wood rot, a strong indicator that bacteria, virus and mold will begin to take over the space. In this case, spinal fluid and cells are the target.

While many of the things I share are solely of my own experience, there are several observations I have made through the years such as grandma had it, then mom or dad had it, etc. The list of obvious symptoms related to fluid contamination I have become aware of are as follows: postnasal drip; ear infection; lymph stagnation or, as I call it in the Eating Yourself to Death book, muddy water; tonsilitis; disruption in thyroid function; kidney disease or pain; bladder leakage; tooth infection; swollen limbs or feet; and frequent urination are all signs for this internal flood of contaminated water. Not only do we have too much water moving around in places it should not be or pooling in unwanted areas, but the water is dirty. The few symptoms listed above are only the symptoms I have identified. What about the fluid issues that are much deeper on the interior of the body and not quite so obvious? What about the fluid that may be collecting in the head and interfering with the electrical activity of the brain? If there is dirty water around the brain, then you can rest assured the spinal fluid is comprised of dirty water. Can this dirty water be the initiation of spinal meningitis? I'm certain the answer to these questions is, "no doubt."

All of this brings to my mind the story of Noah where the flood waters took over the entire earth's surface, only this time we are witnessing a mass of people suffering from an excess of dirty fluid on the interior of the physical body. Is this interior fluid related to "wickedness" as was the cause of the flood in the days of Noah? Well, to a degree, yes. The word wicked is a reference to acts that fall outside of the cosmic laws, either a lack of adherence to or abuse thereof. Many have read them in the Old Testament of the Holy Bible yet to some degree they can be quite vague. They are the laws that God warns if not properly followed judgment (consequences) will ensue. The Bible calls them judgments, today we call them diagnoses. Often the judgments are a result of our ancestors' actions or our own actions, or both. Those who were not following the cosmic laws, the laws that govern nature and the physical body, were classified as "wicked." Fairly stated, we are witnessing the consequences by a means of health issues, and we are still in the dark as to what must be done to escape the ever-continuing judgments that plague mankind. The cosmic laws, when incorporated into daily living, keep the body in proper operation. The knowledge of those laws was swept away through the sands of time by various events throughout history. My assignment is, and has been, to rediscover those laws and figure out how to apply them to daily living that will benefit the health and welfare of mankind.

Hosea 4:6: My people are destroyed for lack of knowledge. Because you have rejected knowledge, I also will reject you from being priest for Me; because you have forgotten the law of your God, I also will forget your children. (NKJV)

The Wisdom of Solomon, Chapter 13 The Folly of Idolatry, Verse 18: He calls upon something weak for health, prays to something dead for life, begs for help from something powerless, and for a safe journey, asks help from something that cannot move.

The cosmic laws must be recovered. Thankfully some of the ancient texts that give us a little better understanding of what those laws are, are being resurrected. The tricky part is interpreting the ancient inscriptions as they apply to the function of the physical body.

As you will see through the information shared in this book, clothing color, fabric and style all play a role in the response our body has to the cosmic activity. We must learn how to build an Ark through our clothing choices that will protect us from physical disease and decay. Clothing plays an imporant role (as well as food) in the structure of the Ark that must be built to save us from the ever-increasing level of dirty water!

Over The Rainbow

This book will walk you over the rainbow of colors, dash you through some rules of style and run through the damages that are often caused by those chemical based materials used to make clothing. Hopefully, a pot of gold will be waiting for all by the final Chapter!

*Clothing gives a voice command
to the interior of the body*

CHAPTER I

THE LANGUAGE OF CLOTHING

Matthew 11:8: *What then did you go out* (into the wilderness) *to see? A man dressed in soft clothes? Look, those who wear soft clothes are in kings' palaces.* (HCS) (description added)

Zechariah 3:4: *So the Angel of the Lord spoke to those standing before Him, "Take off his filthy clothes!" Then He said to him, "See, I have removed your guilt* (sin) *from you, and I will clothe you with splendid robes."* (HCS) (description added)

Scripture describes guilt as filthy clothes or one who is dressed in rags. To assist the body in processing and removing the damaged DNA (sin) that is present in the cells, a proper wardrobe is a must. Your choice of clothing, including color, fabric and style speaks volumes and will attract or deflect the chemical elements that surround you. A person who is improperly dressed is living within the wilderness of worldly systems and exposing their health to numerous unseen, and sometimes mysterious objects. Those who desire to live in "kings' palaces" must dress as though they live in a king's palace. In Scripture, clothing is often a representation of the status of the interior health of the body, most often the plasma (water) and blood. This is a true indicator that your clothing plays a role in the health of the blood, the foundation of health.

Unfamiliar Tongue

Much is said through a person's clothing. Impressions are made within seconds and without the need for sharing a word. Status is reflected through clothing choices and clothing can reflect one's profession. Color, fabric and style choices and accessories all have their own language. What message is your choice of attire for the day sending to those you meet? More importantly, what messages are being sent to the interior of your body by your clothing?

There are some interesting hidden messages that take place between our clothing and our body. Clothing color has been key in the ethnic divisions we see in physical features, skin pigmentation, eye and hair color. The flashy, bright colors, stripes and artistic designs are all contributing factors to the divisions created.

Clothing selections plant seeds that produce visible change in the physical body that have trickled along ancestral lines. For example, through generations of time, wearing genuine furs will produce genetics for a small mouth and straight teeth. Most wearable genuine furs are from animals that have small mouths, and I serious doubt they have crooked teeth. Yes, this all sounds so out of the ordinary yet considering everything has or is a vibration adds a dash of sense to the entire thought. Vibrations record in water and the physical body is approximately 50% water, some have a little more, some a little less.

Colors attract the chemical elements or push them away. Natural fabrics assist the body in erasing any negative vibes we encounter daily. Style can enhance an area of the body and cause that area to become larger or smaller. Is it true to state that the color, fabric and style of your clothing can determine your body shape, weight and features?

I refer to fabric as the collective group of all natural cloth made from goat, sheep, alpaca, cotton, flax (linen) and furs. Material is the collective group of manmade products that contain plastics or other unnatural fibers or chemicals that are processed to become used as clothing, which includes such things as Polyester, Nylon, Rayon, Viscose, Tencel and Bamboo. I do not purchase, nor will I wear any form of recycled fabric or material and the explanation for this will be revealed as you progress through the Chapters herein.

On a scale of 1-5, clothing would qualify for a rank of 4 in importance when it comes to your health. Clothing will either add to the vitality of the Soul (the Star Dust that should dominate the plasma and cells) or take away from it like a debit and credit system. I call this Soul Banking and will address Soul Banking in the future. An easy way to understand this is clothing, like foods, will either increase, protect or harm the light (Star Dust) that is carried by the plasma and housed in the cells. This concept may shed light on why in years past it was common for a person to support a dress and heels, or suit with a tie, that many times was accompanied by a hat. They knew the importance of the quality and style of their clothing and were protecting or adding to their Star Dust. If given some thought, this will explain what baggy or hole ridden clothing, sweatpants, gym clothing and the like have done to the shape of the body. And the Big Dogs of physical exercise had everyone convinced that the ability to stay in shape or be healthy was rooted in burning calories, nutrition and toning exercise!

All interesting, yet how does this all work? I will answer some of the questions behind all this with a notation like many times before, much is still yet to be learned.

Fashion Trends

Fashion trends rotate through the years all in the name of the same subject that has been a repeating strong point, money.

<u>The Wisdom of Solomon, Chapter 14 The Dangers of Idolatry, verse 2</u>: *It is the desire for profit that made the ship, and a craftsman built it using his skill.*

How are those who design and make clothing going to make money if the trends aren't constantly changing, giving rise to the need to restock your wardrobe? Some of the popular fashions have contributed to many a downfall in health and shape of the physical body. A few extra pounds here, and a reduced or enlarged bustline there and we have physical bodies in all shapes and sizes that hours at the gym do not seem to budge. Who would have thought these various sizes and shapes could be a result of the contribution received through clothing? Wardrobes do not have to become drab to avoid the physical shape train-wreck. Ladies, when mature, it is possible to dress youthful, yet tasteful.

Be cautious of reports that should be ignored when it comes to clothing. Many things are hidden within those woven fibers and styles. Not everything the marketing industry reports as being beneficial for the body lives up to what is claimed. "Cooling" clothing and flame-retardant clothing can be hazardous to the blood. The goal is to keep the Star Dust (light) inside the cells alive and healthy.

Read the fine details and do your research before your purchase. An inexpensive article of clothing may not turn out to be so inexpensive if it alters your health. Synthetic materials can throw the electrical system inside the body off course. Not to mention the skin rashes and bursts of sweat that seem to accompany them as well. Please, for the sake of the children, do not dress children in synthetic material nor allow them to sleep amongst them.

CHAPTER II

NOTHING BUT THE TRUTH

Is there an industry, meaning advertising, fashion or design, even food related industry, that tells the whole truth about their product? I question this simply because if people knew the whole truth they might make healthier decisions. Either the education about fashion (or food) is missing or a lack of interest and time to do the required homework exists. The ever-present industry temptations are tossed in the wind and people grab it up like an animal pouncing on its prey.

Half-truths, no truth and a lack of care to research the truth has spilled out manmade material we put on our body that has taken up the title of clothing. There is a flair of artistic talent behind some of the colors, fabrics or styles worn but do we really know what those uneven hemlines, stripes or herringbone prints, tie-dyed and multi-colored, wrinkled or crinkled clothes are doing to the electrical signals that are designed to keep the body in good operating order?

For those familiar with Scripture, the term whore is a reflection of the advertisement and distribution of a product. It is a descriptive word for merchants who go about "selling their goods" whether it be by catalogs, internet advertisements or store displays. An element of physical satisfaction or stimulation is at play here as well. Germanic root of the word whore: a person who desires. European definition: loved, to wish or desire. The fashion industry has done a fine job of initiating desire and many times for items that are not in the least bit beneficial for the health of the body. This brings up the old saying: Be careful what you wish for (or desire). Most people have some knowledge of what can develop from a visit to a whore house! Like a visit to the whorehouse, your clothing can produce consequences depending upon the quality and style you purchase and choose to wear. Use caution when making a purchase.

With all of this in mind, let's step through some not so common territory and see what we can find.

<u>Electric Highways</u>

What needs to be understood is the numerous electrical highways called meridians within the body are influenced by color, fabric, style, and much more. When there is a conflicting object that encounters an electrical highway, a symptom or resulting discord will occur. This is when things such as irregular temperature regulation in the body takes place, or excess water accumulates within the body and the list of symptoms goes on. It is as though an electrical short or outage occurs. These things are not measured within any standard procedure of Western Medicine diagnosis, to my knowledge. The practice of acupuncture will address electrical highway interruptions but will likely not know the root cause.

This busy roadmap of intricate electrical wiring runs the programs our body needs for digestion, cleansing, mental processing, thyroid function and much, much more. Many people may be unaware that what they expose themselves to, what they wear and even what they eat influences these electrical highways, many being located in the shoulders, neck and head region on the body. There is a plethora of gallbladder, stomach, liver and bladder meridian junctures in the head making it vitally important to protect the head from intense or lengthy exposure to the sun, or moon for that matter. Around the outer edge of the ear is a meridian highway called Triple Warmer. When the Chapters to follow reference heat, not being a professionally trained acupuncturist, I can only suspect that "heat" issues in the body can reflect an issue with the Triple Warmer highways contained in the head. Heat can also be an indication of inflammation or thyroid disfunction. One vein of this Triple Warmer highway runs along the outside of the arm to the fourth, or ring finger. It would be quite a different concept to simply adjust the placement or quality of jewelry located at the ring finger in order to calm a heat flush in the body. Maybe it's not hormones causing those heat flushes, but the costume jewelry or style of ring worn. Interesting, nonetheless. Until an expert in the topic steps forward, we move through the hurdles of health issues as best we can, adjusting the seemingly simple items until a symptom subsides. There may be more to the simple adjustments than what one might want to accept.

Many fabric prints, colors, synthetic materials, and styles hinder or deplete the electricity in the meridians and within the plasma, preventing energy highways from functioning properly. Picture in your mind the chakras (wheels) that run through the center of the body. If those wheels are to remain in proper working order, it takes a dose of natural color to help keep that natural rotation of the wheel going. Synthetic materials will slow or prevent the wheels from spinning at the proper rpm. Like adding diesel to a gasoline engine, something will eventually break down. There are enough harmful signals and electronic waves of energy coming at us from cell phones, power lines and numerous other things each day without adding to the mix by our poor choice of clothing. We need our clothing to give us a boost, not drag us into the depths.

··   ··

Activity takes place inside the body that many times we are unaware of until that activity produces an exterior sign or symptom. We must learn what takes place inside the body in response to our clothing choices to hopefully avoid a percentage of the many unwanted health issues witnessed and experienced today. The function inside the body can do some very strange things and learning how all those things work is the challenge we need to conquer.

Fun Fact: Our blood is influenced by the clothing worn by our ancestors as far back as 200 years. The signals from the clothing they wore become recorded in the DNA and contribute to the size and shape of our body.

CHAPTER III
CRAFTWORK

<u>II Chronicles 2:11-14</u>: *Then King Hiram of Tyre wrote a letter and sent it to Solomon: Because the Lord loves His people, He set you over them as king. Hiram also said: May the Lord God of Israel, who made the heavens and the earth, be praised! He gave King David a wise son with insight and understanding, who will build a temple for the Lord and a royal palace for himself. I have now sent Huramabi, a skillful man who has understanding. He is the son of a woman from the daughters of Dan. His father is a man of Tyre. He knows how to work with gold, silver, bronze, iron, stone, and wood, with purple, blue, crimson yarn and fine linen. He knows how to do all kinds of engraving and to execute any design that may be given him. I have sent him to be with your craftsmen and the craftsmen of the lord, your father David. (HCS)*

Before moving forward, it will be beneficial to unravel what the verses above are saying. Foremost, to build a temple is reference to the physical body – "my body is the temple" so, we are not talking about an actual structure of wood or concrete here.

*<u>Solomon</u> means: wisdom

*<u>Hiram</u> means: high-born; exalted.

*<u>Dan</u> has been said to be the "lost tribe" and as the interpretation of these verses unfolds you might agree with this conclusion. The etymology description for Dan is house or household. Certainly, we witness today a household (collective group) of lost (lack of knowledge) people.

*The etymology description for <u>Huramabi</u> is "most noble."

Thus far we know from these descriptions that there is a specific group of people (house), not necessarily "high-born" as we would think of it but of an uncommon DNA we'll say, that construct their bodily temple with a means of metal, wood, fabric and colors. Of note, metals can be made of ore, and ore can give off gases. I'm no expert here so I cannot state that all metals contain ore that emit a gas, to a detectible level that is. This household of people certainly has been lost through the sands of time for the simple fact that the instructions on how to build and maintain this "house" has been lost. We are currently learning what color can and will do. And as for wood, it represents photosynthesis, the act of conversion we know to be done by trees. Point being, there are chemical elements, the Heavenly Gases, involved. Metal(s) is symbolic of precious metals, gold or silver jewelry.

Certainly, the ways for building or keeping the temple, we call "body" in working order and healthy has been lost. How does being a craftsman fit into all this?

Is there an element of craft within the selection of clothing and accessories we put together on any given day? If so, then the same concept would apply to the collection of foods called meals, or colors displayed throughout a home. It is as simple as putting specific items together to produce a desired result.

A quick mental review of words that are used in conjunction with craft are: arts and crafts; craft fair (or show); craftsmen, and the ancient word, witchcraft.

Author's definition of craft: a collection of items put into a specific order or grouping that produce an influence or reaction.

Dictionary definition of craft: exercise skill in making something; a skill used in deceiving others.

I will state that I do not agree with the use of the word "deceiving" in the dictionary definition. A craft may bring change or be a representation of something not fully understood but in goodness it is not intended to be deceptive, it is simply a mystery to logical processing.

The burst of events that came through what we call the witch trials cast a dark shadow on what craft truly is or can be. Officials of the time laid a foundation that has been a muddy mess ever since.

If given some thought, the realization that marketing industries use of a craft comes to light. To entice the sale of a product with decorated displays, background music, verbal promises, and potential discounts, are all a disguise to persuade or influence the sale of their product. A craft is used to bring about a reaction. This suggests, "for every action there is a reaction." Not every craft has a goal of being harmful, it is simply a descriptive word for how putting various items (food, clothing, etc.) together will result in either a beneficial, harmful or neutral signal, a signal being indicative of something that is not yet seen but has an influence.

Keeping the descriptions of craft in mind will aid in understanding how clothing influences the shape, size, color and health of the physical body, albeit sometimes seemingly quite slowly! This is a form of a craft, a set of things coming together that bring change. The Heavenly Gases will respond to the craft it is presented.

When all described herein is considered, one could conclude that every day, to some measure we all participate in a craft. To obtain the internal power needed for living a disease-free life, you must wear all natural fabrics and obey color and style regulations. This is the assigned craft.

ACT I: COLOR

CHAPTER IV

CHEMICAL ELEMENTS TO HEAVENLY GASES

Ezekiel 16:16: *You took some of your garments and adorned multicolored high places for yourself and played the harlot on them. Such things should not happen, nor be.* (NKJV)

Colors attract the chemical elements from the cosmic atmosphere. Once in the body, the chemical elements become what I call Heavenly Gases. The Heavenly Gases produce the Spirit Energy activity. You could say this action of absorption or transformation to Heavenly Gases and circulation through the body is the embodiment or life in the Spirit. Colors are a means for Heavenly Gas transportation, meaning the movement of the essence that originates from the color throughout the body. To a degree, that movement that is required for health is guided by cosmic activity. This is where the moon or even seasons have an influence on the momentum of a color. What does this mean? A color may have a stronger influence in Spring than it does in Winter, and so on. Color is the attractant for the chemical elements that become the Heavenly Gases that initiate the movement of blood, muscles, lymph fluid, spinal fluid, etc. When there is no color, the body has no or at most very little, synergetic fuel source.

Color emits an extension of a hue-like vapor or essence from its particles of origin. For example: I wear a yellow dress and the essence produced by yellow will produce a foggy appearing glow around my body that extends out a few inches from every area covered by the dress, an aura. This aura is not something visible to the naked eye, or to most people it's not. This color induced aura plays a role in the cell messaging that takes place within the body. When the Star Dust cannot move properly, or the body lacks the proper charge from color, mental disabilities or retardation becomes a risk for descendants. Rainbow colors assist cell communication transactions. Ezekiel, in the verse above, warns of the over-use or misuse of multiple colors. This over-use is seen in garments made of numerous colors and patterns that can be bright and busy, which invites an energetic reaction that will come with an internal price to be paid. This price could manifest in the form of physical pain in certain areas of the body or go as far as causing internal disruptions, tumors or genetic disorders.

Not-So-Fun Fact: Sodium hypochlorite (such as Clorox) will remove the Star Dust light within the body just like it removes color from fabrics.

CHAPTER V
SEASONAL COLORS

Pastels are the most beneficial by not being overpowering in their message. Dark colors are permitted in certain situations, but I would advise against wearing dark colors on a daily basis. Dark colors like that of denim or navy cause the internal temperature of the body to become too cool to host the Heavenly Gases. While dark colors attract heat to the exterior of the body, they cool down the internal temperature that controls the livelihood of the Heavenly Gases. It takes a lot of internal power to wear dark or overpowering colors every day and draining the internal power is certainly not beneficial, likely leading to health disruptions. With this in mind, a company that requires their employees to wear dark colored uniforms might consider and offer light(er) color selections or a rotation of color for uniforms.

The measure of influence a color can have is determined in part by the season in which it is worn. Each year may present some slight variations to the influence of a seasonal color. This seemingly confusing system of color influence places the decision of the seasonal and yearly color selections solely in God's hands.

Spring: yellow, teal, pale orange shades, lavender, pink, blue. Small flower prints permitted.

Summer: teal, yellow and yellow gold, lavender, pink, blue. Wear solid colors, no prints. Lavender and aqua are cooling shades and will assist in cooling the body during hot Summer days. If the body has temperature regulation interruptions, this can be genetic, from eating spicey foods over an extended time, or thyroid disruption. Wearing lavender will assist the body in resetting the temperature regulation the body is to naturally have.

Autumn: green, yellow, yellow gold, orange, red, camel or tan. Small floral prints are permitted.

Winter: green, red, blue, purple, pink, camel or tan. Wear solid colors, no prints. Green attracts oxygen and is beneficial during the cold months. Green grass and trees take care of this job in Spring and Summer.

No doubt, colors influence the blood. This influence can be in a beneficial form by correcting certain residual essence vibrations left behind by such things as food coloring, additives or preservatives, etc., or a harmful form through encounters with chemicals. Food combinations can be a recipe for gases that damage cells or disrupt the beneficial gases. Color can help the body correct these things.

Mix-matched clothing colors cause a heat issue in the body. Shades of one particular color worn at any time are more beneficial than wearing a display of numerous color combinations.

CHAPTER VI
WHAT IS THAT COLOR SAYING?

In the previous chapter I shared one of the benefits of the colors lavender and aqua. In this chapter I share what I have learned about various colors with the caveat that there is still much to be discovered on the impact color has on our health. As time moves forward and younger generations rise up, information will continue to unfold.

Psalm 45:13-14: *The royal daughter is all glorious within the palace; her clothing is woven with gold. She shall be brought to the King in robes of many colors; the virgins, her companions who follow her, shall be brought to You.* (NKJV)

Everything has a language, including color. According to the verses above, if we want to "see the King" we best be dressed in the appropriate colors. The color palette has become quite large giving us many options to select from, some of which are beneficial, and some are, well, not so beneficial. The basic pastel colors are the most important with the variation in hue having the ability to increase or decrease and still be in a beneficial range.

Colors influence the mRNA process. This activity may have a connection to a color that is enhanced within its specific season. As stated, clothing fabric and color are huge players in the conduction of electricity in the body. Let's sift through the story of Joseph with his colorful robe and see what unfolds.

Genesis 37:3, 31-32: *Now Israel* (Jacob) *loved Joseph more than his other sons because Joseph was a son born to him in his old age, and he made a long-sleeved robe for him.* Verses 31-32: *So they* (Joseph's jealous brothers) *took Joseph's robe, slaughtered a male goat, and dipped the robe in its blood. They sent the long-sleeved robe to their father and said, "We found this. Examine it. Is it your son's robe or not?"* (CSB) (descriptions added)

If Joseph stood out within the crowd of his eleven brothers because of a colorful robe, it is easy to conclude something important goes on within the body that produces and responds to color. Joseph, being favored by his father because he was the son of the favored wife, Rachel and came about during the later years of Jacob's life, has a DNA component at play on the inside that reflects through to Joseph's exterior. When Scripture speaks of an exterior cloak or coat it is a reference to activity that is going on inside the body that manifests in the form of maybe a gift of clairvoyance, intuition or a skill on the outside of the body. The manifestation is not always something that can be seen by the physical eyes but rather a nature about the person. In Joseph's case, his interior activity was creating an aura in a rainbow of colors, a result of the connection to the cosmos, originating in the depths of his DNA. That aura was a contributor to Joseph's ability to have memorable dreams. For whatever mystical reason, Joseph had cosmic related gifts that none of his brothers had or recognized, and they despised Joseph for it. In the verses above, notice the reference to blood and the connection to the robe.

••   ••

There is a reason, possibly many, why the rainbow seen after a shower of rain appears in the sky to display its array of beautiful colors. We should surround ourselves in pastel or gentle hues, nothing dark or overpowering, no black, gray, navy or white. As a simple reminder: Follow the rainbow.

When the body experiences a setback in health or a surge in the normal level of energy, the color of clothing you select for the day can either hinder or help the physical symptoms being experienced. Similarly, the color of paint on walls in any office or home will produce an energetic influence on your mood, physical energy and blood.

During the period when I was experiencing a significant health setback, I had the interior walls of my home painted, replacing the soft gray with a light golden yellow. Any room with gray tile, marble or quartz received a splash of lavender color to offset the gray. It would be impossible to measure the level of change that took place with the new color scheme, but I did notice a sense of calming to my body. It is quite possible this simple change removed a burden and gave my body a more suitable opportunity to address the health issues at hand. It is well reported that bedroom wall color can influence quality of sleep. Clothing color is very important, although the impact is often more subtle than an entire room of color.

It's All Black and White

White has its place but .... not on your body or on your bed. Well, let's back that up a bit. White is what should be worn with respect to undergarments, the undershirt, panty or boxer, bra, socks. For the exterior wardrobe, white should take a backseat. White interferes with the natural cleansing process that takes place inside the body. What do I mean by this? Fluids in the body produce a pulsing style cleanse cycle. Imagine the function of the lymph system, and most are aware that the blood has a cleansing process to it, then what about the spinal fluid? This cycle of cleaning is enhanced on specific days of the week. When white is worn it interferes with the electrical energy it takes to run this natural cycle. This could be compared to loading your washing machine, adding the detergent but having no electrical power to run the cycle. White cancels the necessary signals to get the job done. You could say white puts the brakes on the natural flow necessary to keep the body energy moving in all its proper directions.

White influences the Heavenly Gas Argon (and possibly others) and will throw off the yin/yang balance. Wearing white daily causes stagnation of energy to a point the energy will begin to pool in one specific location causing weight gain in that area. For example, if you wore a white blouse with green capri pants, the natural energy flow of the body would become restricted to the lower portion of your body where the color green is. If this pattern of color choices continues, after a period of time weight would accumulate in the lower region of the body. Where the energy is, the body will reflect that energy in one way or another. A white, pressed business shirt is necessary at times but without a doubt, colored shirts better serve the energy aspects. Men, wear a colorful tie and a suit jacket with your crisp, white dress shirts.

Not-So-Fun Fact: A white wedding dress and white pants both contribute to the enlargement of the derriere!

If white is this powerful when we wear it, what are white bed linens doing to our body as we sleep? There is discovered and undiscovered activity that takes place inside the body as we sleep. An internal clock chart will show some of this activity. There are additional activities that take place inside the body involving the Heavenly Gases during nighttime hours. Color is needed for those activities. Would someone please notify the hotel industry about the importance of color when it comes to sleeping. The white linens have to go! Both the moon and the sun distribute chemical elements we must be in proper position (including color) to receive.

Wearing black and white was made big by the fashion industry and snatched up by many within the circle of the business industry. The fashion industry, like many other industries, does not take into consideration the electrical influence color combinations or prints have on the body. What happens when we wear black with white? Swollen feet are one symptom. Mechanisms involving involuntary rotations of the body are another. This could manifest in the rotation of the neck and head, ankle movement or pivoting of the wrist. Black and white prints influence eye color. Eye color, in part, reflects the frequency of light contained in cells.

Black prevents the plasma and spinal fluid from producing its natural song I spoke of previously. Healthy vibrations come together as the fluids flow in proper proportion and rhythm and a form of humming will take place. I compare this to the song that is heard in the galaxy. The energy is so great in measure that the flow of forces allows it to create a mesmerizing sound. When the plasma and/or spinal fluid lose the electrical charge, the song will not be present, and the body will suffer physically. Think of this like an electrical power outage in your home. No power, no music.

Black panties, trousers and skirts target the region of the belly and will contribute to the collapse of the energy field within you. This can result in digestive issues and many other interruptions in organ function within the region of the belly.

Black shirts can cause you to look older than your birthday may reflect. Black leggings will do the same. We all know that black absorbs heat, but did we know that the heat absorbed through black can become stuck within the body causing the natural temperature regulation within the body to malfunction? First you may feel too hot, then you may feel too cold. Again, this brings up the question of whether the thyroid function is influenced by black since the thyroid is a temperature regulator.

Wearing black lace will trickle along the generational lines and influence the chest and shoulder region of male descendants. Lace is also a culprit for interfering with the function of the lymph. As tempting as it may be ladies, it would be wise to avoid wearing lace across the chest, particularly for those with a history of breast cancer, or the genetics for it. Wearing black clothing set the stage for black hair.

The legacy of wearing black to a funeral should come to an end. We do not need everyone in attendance at a funeral or memorial to end up damaging their internal light. Think of how refreshing it would be to attend a memorial or funeral, and attendees were dressed in color, creating a rainbow effect on the atmosphere. Who knows, that display of color may help the deceased complete their journey to their final resting place.

The important thing to remember here is black, white and black and white together interfere with the healthy function of the electricity within cells. The only place that black and white are to be worn together is on your feet. White cotton socks that cover the ankle with black leather shoes.

Then Came Gray

When did gray come on the scene? The word gray (and grey) dates back to 700 BCE and holds a definition of: to age, wither, become gray. Sounds like something I wouldn't want to associate with. Gray came about through undyed wool and was worn by peasants, servants and religious orders. How did it become so popular with the wealthy? It only takes one person who lives in the spotlight to begin a ripple of change. When that person in the spotlight begins to support clothing in shades of gray the general population soon follows.

What does all of this say for gray? The color gray is one of the culprits for activating improper fluid issues within the body. There seems to be, and it is sometimes challenging to put your finger on the exact issue, a collection of colors, foods or simply being exposed to plants within the grass family, that trigger an influence directed at fluid in the body and it ends up in places it shouldn't be or causes it to become dirty. The process of how this takes place is yet to be discovered. What I have discovered are some of the triggers. Gray is guilty of contributing to the signs of aging. This aging process comes through the condition of the fluid in the body. What should be in the body is a clean running stream you might find in the mountains, clear and in its proper place. For the body to have the ability to carry the electrical charge necessary for proper organ function, the plasma, spinal fluid, and cellular fluid must remain free of debris.

The fact that gray is "not black and not white" brings to mind the Scripture below:

Revelation 3:16: *So, because you are lukewarm, and neither hot nor cold, I am going to vomit you out of my mouth.* (CSB)

The symptoms discussed may all seem like small issues but tally them up over generations and you come up with genetics that produce disease. A collection of signals that do not align with the origin of DNA's design and eventually the burden is too much, causing the cellular structure to become stressed.

Fun Fact: Gray is a color that is disruptive to any anti-aging attempts.

The History of Purple

Since ancient times, purple has been associated with Royalty. To dye cloth into a purple hue was expensive, making the purple cloth a bit pricier than the average person could afford.

Acts 16:14: *A God-fearing woman named Lydia, a dealer in purple cloth from the city of Thyatira* (lunacy), *was listening. The Lord opened her heart to respond to what Paul was saying.* (CSB) (description added)

Mark 15:17: *They dressed him* (Jesus) *in a purple robe, twisted together a crown of thorns, and put it on him.* (CSB) (description added)

The entirety of the story of Jesus is about the influence on, and condition of the blood, plasma, and spinal fluid that carries the DNA. Mark 15:17 tells us that purple has an influence on the blood (and spinal fluid) that focuses on the brain at the crown of the head.

Lavender and purple are uplifting shades, and I have not encountered any negative results from either. Purple shades support brain health and help erase things such as residue left in the body from the childhood snacks like sugary popcorn like Cracker Jacks or kettle corn, and quite possibly many more of those sugar-coated snacks children enjoy. Odd, but true.

Purple supports floral prints well. Just try to select the prints with small flowers lest certain areas of the body will "bloom", and not necessarily in ways you would desire. Purple helps reduce the dirty water issues that the color gray seems to activate. The lavender to purple shades assist the body in erasing harm done to the cells from exercise. As mentioned in previous publications, exercise is not the same as remaining active. Activities that cause an increase in the respiratory or heart rate for an extended period of time classifies as exercise and consequently depletes the Heavenly Gases from the cells.

Try to avoid shades in the plum category. Plum will produce descendants with a broad derriere. How is it that many of the troublesome articles of clothing, and color seem to afflict the size of the bum?! So much for the assumption that you have simply eaten too many carbs.

Fun Fact: When the design and style of kitchens takes on a new identity, lavender and aqua are good cooling colors and work well in kitchens.

The Blues

Numbers 4:7, 9: *They are to spread a blue cloth over the table of the Presence and place the plates and cups on it, as well as the bowls and pitchers for the drink offering.*

The regular bread offering is to be on it. 9: They are to take a blue cloth and cover the lampstand used for light with its lamps, snuffers and firepans, as well as its jars of oil by which they service it. (CSB)

The verses in Numbers that speak of the blue cloth are referencing internal activity, meaning within the physical body, not necessarily a religious building. It appears blue dances well with Hydrogen that would come from the bread in verse 7. Lamps and light are a reference to receiving to a point of comprehending. In this case, light that is produced by the presence of Hydrogen. This light would be in the cells. With this connection, I conclude that shades of blue (baby blue to sapphire or cobalt) play a role in intelligence which is a result of a healthy forebrain. The reference to table puts the focus on the/a current situation that is dependent upon a form of intake (information or, food); cup is the gift of divination as told in the story of Joseph and Benjamin (Genesis chapters 40 and 44), having a knowledge of events hidden or yet to unfold; bowl is where things are held or collect like the head/brain/intellect; plates hold items that can be challenging to chew up or swallow (accept), and food that is partaken of. Light that should be in our plasma and spinal fluid is influenced by blue. When all things are in working order in the body, the plasma will be the River of Light that flows within us. The color blue obviously plays a role connected to this light. In any case, blue is the color that should be used to help straighten things out.

If blues are a favorite, light shades such as baby blue produce the most benefit. Sapphire is an excellent choice as well. Avoid the deeper, darker shades like navy, including those often found in denim, and blue grey tones. Blue is a water signaling color and will have an influence on the lymph and other water processes within the body. These signals, dependent upon an individual's situation, can be of benefit or cause disturbance to the bladder and/or kidneys. Blue sends signals to the fluid and thus should not be accompanied by white. This combination can encourage increase in white blood cells. Wearing blue with white can also interfere with the meridians that govern the function of the heart. In its season, light blue shades assist the mRNA process. Blue has an influence on the ability to regulate sleep-wake cycles known as circadian rhythm.

Navy blue is overpowering for the body and borders on the shades of black. Formals or long dresses that are predominately navy will cause weight gain in the belly, thighs and hips; is a contributor to olive colored skin and has an influence in the region of the clavicle. Navy can also have an influence the crown of the head. Certainly, navy blue should not be worn in any form from the waist down. The only exception to this would be navy shoes that can be worn with light blue or mauve colored clothing. Darker blue colors can cause you to be spiritually vulnerable in the form of being exposed to someone's energetic baggage. Wearing blue and white pajamas can result in emotional and literal heart break. Blue and white stripe contributes to loss of the front/back identification of the DNA strands. Personally, I avoid all stripe patterns.

Dark colors across the board will decrease the production and/or function of enzymes. Blood temperature is also more easily influenced by dark colors. Does the body use the enzymes to balance out the statement the dark colors give off? At this point, that remains a mystery. Those darker shades are simply just too challenging for the body to handle without a compromise.

Earth Tones

Wearing brown can cause the skin pigmentation to become tan/darker. This is particularly true when wearing leather clothing or coats. Who would have guessed the color of your skin could be altered by the choices of color worn? This subject can be put into the category of Universal language that is not always understood.

If you wear tan or light brown, accompany it with a pastel. These tan, light brown and camel shades use up more of your energy than lighter shades so adding a bit of color to the mix will help maintain balanced energy movement.

Falling for Green and Teal

Autumn and Winter (specifically October – March/Spring Equinox) is the time for green. Contrary to some popular fashion trends, Spring is not the best time for wearing green. Green accompanied by pink floral print will initiate issues that give rise to medical related conditions. This would be an internal issue related to the activity or measure of gases in the body. Green influences the mRNA process. Green is a participant in photosynthesis, putting green in a category of a transformation and transfer color. Green opens the opportunity to give and receive cellular signals.

The color green will attract Oxygen and would therefore act as a benefit when eating at a restaurant where Oxygen levels can become depleted in response to the heat production from the kitchen. Another option would be to wear Emeralds or carry an Emerald, Beryl or Jade stone in your pocket. Green can also attract Nitrogen, and the Nitrogen is a player in the copying or transfer of cellular information.

Green will clear out any unwanted influence from animals/pets. Those loveable, furry friends we have, and for me that is either a cat or a dog, all leave an invisible influence on our body. To keep things in check, wear green during its proper season to erase any potentially draining pet transferred signals.

Use caution when painting walls in a kitchen area green. Activity within the walls of a kitchen that are green can result in an internal muddy water situation causing the lymph to become overloaded and sluggish and plasma light to decline. With green being a transfer color and being in the kitchen, there is a potential for signals from unwashed produce or blood from uncooked meats to transfer to the cook resulting in unwanted and potentially harmful signals recording in the fluid inside the body.

Teal assists in cleaning the lymph. Teal also aids in the elimination of plastics from the body. This might be a simple, yet beneficial therapy for small children. This is not to say the entirety of plastic toxicity will be eliminated by wearing teal colored clothing but there is a strong indication teal aids the body in the processes necessary for the removal of plastics.

Color is involved in moving cell messages in the body with blue and green being prominent in this process.

Psalm 23:2-3: *He lets me lie down in green pastures; He leads me beside quiet waters. He renews my life; He leads me along the right paths for His name's sake.* (HCS)

In this verse, there appears to be a green connection to calm(ing) waters. This would be the influence of the color green on the fluid in the body. This calm water renews or heals various issues in the body. Those issues would be things one might see in a blood panel like cholesterol, gases, nutrients, and so forth. The calm water could be said to settle inner chaos.

The interplay between green and the blood (represented by the color red) is displayed in many Christmas themes but a word of caution I must throw out. Green is a player in the mRNA activity and the color red supports the blood. The natural cycle of cell messaging can be interrupted or thrown off course when green and red are worn together. The green and red combination is a culprit in the over production of red blood cells (Erythrocytosis), an issue in my family genetics. Bottom line, avoid wearing a green and red outfit – these two colors do not play well together. The seasons of Autumn and Winter support these two colors but putting them together overpowers the blood resulting in a harmful influence. Wearing green will help balance a high red blood cell count. Overall, it appears green is a multi-tasker!

Fun Fact: Green and white checked fabric will eventually result in big feet.

Pretty in Pink

Need to wake up your intuition? Wear pink. Pink also helps correct nighttime hot flashes. See more information on this in Chapter XI, Time for Bed. Pink is the color that connects to skin health and may be a helpful remedy for skin rash or eczema. I have not had reason to try this but applying a pink cotton cloth to a skin irritation may help in clearing it up, or at the least bring some relief. Pink helps remove harmful energies left behind by artificial colors that are often found in candies and canned beverages.

Pink can make you more susceptible to sharing cell signals with others. This can be either good or bad, depending on who you are in contact with. If your schedule includes being in close company with someone who is known to you as a disruption or irritation, it might be wise to avoid wearing pink.

Hot pink is a strong color and can alter the characteristics of a Spiritual Royal (fair skin, blue eyes, dark brown or chestnut hair).

A Splash of Orange

Orange joins the ranks of hot pink. Does this mean the Denver Bronco jerseys need a new color? That's a debatable topic. Something about the mix of red, the color that triggers a blood response, and yellow together drains the energy. Worn in the proper season of Autumn, it may not have such a draining influence. Orange influence could hinge on a few factors: 1) season is foremost; and 2) individual genetic makeup. Some people may get by with orange but there is an entirely unique group that is altering their DNA strands by wearing orange. Orange-red pulls its energies from the planet Mars.

Scarlet, Crimson and Red

Reds can make such a powerful statement being bold and beautiful. Your clothes, your food, your red car, red paint, all give rise to the importance of using a level of caution when wanting to support red.

Long dresses, specifically the deep red shades like burgundy or wine, join the ranks for afflicting the blood. These rich, deep red tones can result in a need for assistance with general daily activity, producing a crippling type of symptom as life progresses. Any of the red tones must be worn with proper accompanying colors or accessories. Outside of wearing black leather shoes, black should never accompany red.

Red with white or black will deplete the Star Dust in the plasma by either cancelling or blocking the electrical signal that is important for the production of light in the cells.

Isaiah 1:18: *"Come, let's settle this," says the Lord. "Though your sins are scarlet, they will be as white as snow; though they are crimson red, they will be like wool."* (CSB)

Isaiah 1:18 reveals that sin is an issue that takes up residence in the blood (crimson). It also tells us that once those "sins" are eliminated, the blood will change and support a light producing element (white, snow, wool). Wouldn't it be interesting if the blood turned to a milky white once all genetic contaminations were removed? Sins are the genetic imprints that float around in our blood that eventually pop up and make their debut as disease. The Lord describes how those sins will change, not necessarily to white but through light (Star Dust) that is carried by the plasma and spinal fluid. I'm going to jump ahead of my topics, but I must point out that the reference to wool tells us that pure natural wool clothing speaks to the blood in a way that helps erase the sin (genetic imprints).

Sunlit Yellow

Ah, yellow, the color of the sun. Spring and Summer are the best seasons for pale yellow. Yellow has a few powerful beneficial aspects. Yellow gold aids in protecting the body from EMF and proves to be beneficial these days. This protection includes the use of headphones, cell phones and laptops. Yellow gold attracts good vibes and can be put in a category of life-giving. Yellow targets helping remove plastics from the blood and helps with digestion issues. Any genetic imprints or self-infliction of cursing or swearing will become quieted by wearing yellow. Dry lips? Chronic dry lips are a result of the consumption of duck, think of the duckbill. Yellow will help sooth this chronic condition.

Combos

There are colors that play well together and some that do not. Clothing that supports numerous colors will send confusing messages to the body and can drain the energy rather than support it. This would mean clown costumes may need to take a rest.

In Spring, yellow and orange attract the chemical element, Argon, and likely a few of the other elements.

Lime green will interrupt the level of Argon, Oxygen and Nitrogen gases within the body and potentially lead you down a path to the physician's office. Green with peach during Last (3rd) Quarter Moon can result in phantom pain in the spine.

Be cautious of neon colors! The language of neon colors interferes with the Heavenly Gases within our body. Neon colors produce situations where the Heavenly Gases inside the body seem to seep out, like a tire that has a slow leak of its air. Here is another situation that warrants caution and unless necessary, avoidance. The calves of the legs can experience discomfort in response to neon colors. My advice is to educate yourself on neon colors and how they are made before deciding to purchase that neon t-shirt or jacket. The price you pay for such items may be in your health. There is an interesting article at publication@medium.com on the history of neon colors.

Pink with gray is another combination it would be best to avoid. Pink with gray can result in a burden on the shoulders. These two colors together interfere with the benefit the cells can receive during certain moon phases.

The L.A. Lakers may be glad to hear this, purple or lavender with yellow or yellow gold is a benefit to the body. I cannot say this color combo can take credit for the team's skill but who knows? In part, maybe it can.

Navy, red and white stripes are a recipe for a virus with a fever with the root cause being an alchemy disruption. Maybe we should reconsider the colors of the U.S. flag and likewise the celebratory attire worn on July 4. Clothing of tricolors can make you appear older. I'm not convinced that many desire this type of result from their wardrobe. Wearing shades of the same color provides a beneficial result.

What if white is accompanied by red? As previously shared, red is the color that attracts the attention of the blood. Anything worn that is red should not be coupled with black, green or white. Red with white causes decay of the cells; red with black shuts down the power of the cells. Red with green was discussed in the subtitle Falling for Green and Teal.

Combing blue and white, particularly for underwear, will cause bladder leakage. Blue with white will block the ability to receive the Heavenly Gases that are necessary for healthy plasma and production of water (Hydrogen +

Oxygen) within the body. Blue with white also constricts the region of the chest, interfering with the proper function of the meridians.

Wearing blue and orange together will initiate genetics for diabetes. Denim blue bleeds its influence into the sphere of the color it is worn with. For example, if you wear denim or dark blue with pink, the blue will overpower the influence of the pink and you end up with a distorted message given off by the pink. Wearing denim jeans with an orange shirt can result in chronic weight loss becoming recorded in the genetics. The heavy woven strands of dark colored fabric act as a barrier, preventing the body from receiving the necessary chemical elements present in the atmosphere. Eventually, the DNA will suffer in the form of wrinkles (creases or folds) that can form in DNA strands. Denim may be a contributor to that issue. I'm sure there are many more contributors waiting to be discovered.

Color in the proper combinations, during their appropriate moon phase and season can provide a sphere of protection around your body. This is not only a daily benefit, but extremely beneficial during public appearances and travel.

Fun Fact: When health issues seem to be impossible to change, try altering a color scheme, whether that of your clothing or in your bedroom.

CHAPTER VII
WHAT'S THE CHARGE?

Colors have a negative or positive ion charge to them. There are specific colors that benefit or hinder the ion charge a Spiritual Royal must maintain.

Red: negative
Green: positive
Yellow: negative
Purple: positive
Light blue: negative
Denim and dark blues: neither positive nor negative
White: negative
Black: positive
Beige: positive
Pink: negative

Reviewing the information pertaining to negative and positive charges brings the question as to whether the color of clothing inspired blood to evolve into being positive or negative in its charge. If color has this ability, it is quite likely food has this influence as well.

CHAPTER VIII
MOON PHASE
AND CLOTHING COLORS

Colors are not only influential by season but also by the phase of the moon. I have applied wearing certain colors during specific moon phases. As a general guideline, listed below are the color selections I made for a particular moon phase. This selection of colors may or may not work for others. I bring this information to your awareness as an aid for those situations where a health issue just does not seem to want to improve, like being tired or having periods of insomnia, phantom pain or numerous other irritants. As the earth advances toward becoming settled in its magnetic charge and as more experience is had with the influence of color, this chart will likely see some adjustments. As with many things, we must be willing to adjust as nature dictates.

New Moon to 1st Quarter Moon: NO RED; wear blue on day of New Moon; permitted colors for the week: blue, yellow, yellow gold, lavender, pink, green and teal. The month of December brings some alterations to this list of colors. mRNA activity takes place November -January, starting at New Moon following Thanksgiving.

1st Quarter Moon to Full Moon: NO RED; wear green on day of 1st Quarter Moon; permitted colors for the week: blue, lavender, yellow, yellow gold, teal, green.

Full Moon to 3rd Quarter Moon: NO BLUES; wear teal on day of Full Moon; permitted colors for the week: camel, red, yellow, yellow gold, teal, green, pink.

3rd Quarter Moon to New Moon: NO BLUES, NO PASTEL YELLOW; wear red on day of 3rd Quarter Moon; permitted colors for the week: yellow gold, red, teal, camel, pink, lavender, green.

If your body needs a boost in the mRNA process, you could benefit from wearing green OR red between New Moon and Full Moon during the November-January dates mentioned above.

ACT II
FABRIC

CHAPTER IX

SECOND TIME AROUND

Wearing used donated or second-hand clothing can result in some undesired physical alterations. Energy is carried through fabrics and when a particular clothing item has been regularly worn by a person their cellular signals permanently record in the fibers. Those prerecorded signals coming through to a different person now wearing the garment can result in electrical havoc to the body of the latest recipient of the clothing. This energy shift, or transfer of sorts, happens not only when clothes are passed from one child in a family to a sibling or other family member, but also when clothing passes to an unknown party. Could sibling rivalry be a manifestation of an inner irritation from the unseen hand-me-down activity? I'm sure there are many more unidentified hand-me-downs, or secondhand clothing issues yet to be identified.

While purchasing clothing made from high-quality fabrics can become expensive, the value received through the high-quality fabric with respect to health will save spending money to correct physical illness or disease. You get what you pay for, and this is true when it comes to purchasing the garments that will envelop your body day or night.

When possible, I suggest having a tailor make your clothing. First, the clothing will fit your body perfectly and second, the article of clothing has not passed through numerous hands during the production process.

Not-So-Fun Fact: Second-hand clothing is a contributor to excessively large bustlines and alterations in skin pigmentation.

CHAPTER X
NATURAL FABRICS

Fabric has a neutral, positive or, depending on the quality, negative influence on the electrical activity within the body. It can either assist in the health of the chakras (wheels) that reach from head to tail in the body or protect the existing energy that those chakra points hold. All natural, no chemical content or treatment, and organic fabric should be utilized as much as possible. Yes, it can become an expense but as stated in the previous chapter, purchasing and wearing high quality fabrics are a more beneficial investment than spending time and money at the physician's office.

Cotton, preferably organic
Wool
Vicuna (soft wool; native to Peru)
Linen
Alpaca
Cashmere
Silk

Cotton is the go-to fabric for undergarments and night wear. Cotton and linen both serve well in the warmer months. I have found linen bedsheets to be quite comfy in Winter as well as in Summer.

Wool has a level of healing properties that work at erasing certain essence leftovers you have encountered that could become potential health problems. During periods throughout history wool was a common fabric used for daily attire. Remove the wool and replace it with synthetics and look at the collection of health issues that abound. I have never seen Vicuna wool so I do not have any specifics on it, but I have no doubt that it would be of great benefit as well.

An odd thing that most people would never think of is wool helps to remove energetic residue from foods that contribute to inflammation, especially those in the sweets category like cookies or cakes. Batter and dough, which can contain raw eggs, trigger inflammation. This type of activity that removes and corrects residue left by food will produce an interesting subject as more information and resources become available.

Song of Solomon 4:2: *Your teeth are like a flock of newly shorn sheep coming up from washing, each one having a twin, and not one missing.* (HCS)

Job 31:19-20: *...if I have seen anyone dying for lack of clothing or a needy person without a cloak, if he did not bless me while warming himself with the fleece from my sheep...* (HCS)

Wool not only has a naturally superior synergetic component, but it attracts the chemical elements that are needed by our body to erase the residue from hindering or disruptive vibrations. According to Solomon, wool contributes to healthy teeth and Job tells us that the fleece from sheep brings life to the body and internal wealth. The chemical elements work through animals of various categories like they do humans and to wear an article of clothing made from wool will produce positive results in the area of health. Alpaca and cashmere hold the same or similar health benefits as sheep's wool. Wool and sheep are mentioned multiple times in Scripture in varying illustrations so it is no wonder wool plays a valuable role in health.

Look at Psalm 50:9-10: *I will not accept a bull from your household or male goats from your pens, for every animal of the forest is Mine, the cattle on a thousand hills.* (HCS)

What is the verse telling us? Animals that are left to their natural environment, not domesticated or penned up, have a photosynthesis ability. Their hide, fur or wool absorbs chemical elements and those chemical elements transfer through and take up residence in their coat. Having said that, it is also said in Book 2 of the Books of Enoch, 59.5: *But whoever kills beast without wounds, kills his own soul(s) and defiles his own flesh.* This tells us that the killing of any animal is to be left to situations where the animal is wounded (which would include untreatable illness).

To drive a nail into the point that animal fur/coats are an excellent option for regaining and maintaining health, Book 2 of the Books of Enoch, 59.2-3: *For man brings clean animals to make sacrifice for sin, that he may have cure of his soul. And if they bring for sacrifice clean animals, and birds, man has cure, he cures his soul.* To be clear, sacrifice does not mean butcher! Sacrifice means to consider of great value or to give honor. Sacrifice is also a reference to blood. Animal fur that comes from free roaming or a natural environment provide a level of healing to the blood of humans.

Previously referenced in subsection Scarlet, Crimson, Red: Isaiah 1:18: *"Come, let us discuss this," says the Lord. "Though your sins are like scarlet, they will be white as snow; though they* (sins) *are as red as crimson, they will be like wool.* (HCS)

Linen has come to the surface in the past few years with reports on its benefit to the body. A google search on the vibration of fabrics will get you hooked up with some information on the subject. As noted in previous publications, linen was the fabric Jesus was wrapped in upon his death and we all know what the remainder of that story is. Obviously, some kind of good energy going on with the linen.

NEVER wear wool with linen, whether as separate pieces or as a combined fabric. Be cautious of the combined fabrics and check all labels prior to purchase. There are scientific studies on the results of combining linen and wool. A quick google search will provide the information you may be interested in on this subject.

Deuteronomy 22:11: *Do not wear clothes made of both wool and linen.* (HCS)

Alpaca is a favorite of mine. It is not only warm and soft but aids in the production of the rainbow of colored light within the body. The chakras (wheels) inside the body love alpaca. The more light producing the fabric, the more protection from outside invaders is available. An outside invader can be anything from pollution to noise to another person's emotional essence. Alpaca can be expensive so one might opt for scarves, gloves or hats when on a restricted budget.

Cashmere is another favorite. With many varieties of color and style, cashmere is a great option. A nice cashmere sweater feels good on a cold Winter's day. Cashmere is from the belly of the cashmere goat, and originating from the region of the belly, it can't help but pass on a benefit to the digestion and other functions located within the abdomen. If you have digestive issues or are just having a bout of upset tummy, wrap yourself in a cashmere blanket.

Real silk, nothing man-made is permissible. Silk feels nice on the skin and is said to help in the prevention or reduction of wrinkles on the face when used as a pillowcase. Silk wrinkles easily and is best pressed with a steamer. Silk can also water spot quite easily, so it is not a good choice for a day of cooking or housekeeping. May need to reserve the silk for those business meetings or dressy occasions.

Leather will result in the skin pigmentation becoming darker. Leather is permitted when it is used for shoes. In the case of leather, you have skin (of an animal) speaking to skin (of a human) although when wearing a shoe, you have a white sock that cancels the tanning signal that would come through to the skin. This creates a form of dividing line between the body and the feet. Leather can hinder learning skills by interrupting the natural function of the hippocampus. It is advisable to not dress children in leather. Suede leather interferes with the Heavenly Gases.

Furs that would originate from animals that can be pets like rabbits, worn on or around the head can result in ear or brain fluid issues and ringing in the ears. Pet friendly furs are not advised.

Fun Fact: Appropriately press the wrinkles out of your clothing. Wearing wrinkled clothing does what? Produces wrinkles! So much for the clothing that supports the crinkled look!

CHAPTER XI
TIME FOR BED

Proverbs 7:16: *I have decked my bed with coverings of tapestry, with carved works, with fine linen of Egypt.* (KJV)

Bedding is just as important as clothing. Make certain you use natural fabric mattress pads, linens and organic mattresses. Sinus issues can be a result of synthetic bedding, including the mattress. Sinus issues also stem from covering the head with blankets while sleeping. Avoid printed fabrics for bedding. There is much activity that takes place between the planets and our bodies during sleep and we benefit the most when all things active can flow smoothly.

Synthetics, including microfibers, disrupt the electrical activity necessary during sleep to recharge and clean our internal organs. This means those soft, fluffy blankets and foam pillows must be left for the pets. Fluid accumulations in the spine and frequent urination issues are a result of a toxic bedding and bed mattresses. Synthetics may also be guilty of causing childhood bed wetting. Unmade beds and beds with articles of clothing left in them will cause you to be taxed in the Soul Bank account department. The Soul is the light (Star Dust) carried through the veins by the plasma and housed in the cells.

Pink bedding helps restore healthy temperature regulation in the body. Sleeping alone also contributes to the natural balance of the interior temperature of the body. Ladies, if you are challenged with heat flushes, it may not be the hormones but the simple fact of sharing your bed with someone else. Simple fix, (while smiling) kick them out!

Fun Fact: Being zipped into a sleeping bag that 1) is synthetic; and 2) causes restriction of air and energy circulation will result in chronic bad breath.

CHAPTER XII
PLEASE PASS THE TOWELS

Bath towels in dark shades of brown, gray or black will disrupt kidney function and influence the urine production and elimination. This is a good place to interject that cleaning products disrupt the kidneys as well.

Though quite popular, white is not well liked by the body when it comes to light production, bath towels included. White cancels out what the rainbow colors revive. Use towels in pastel shades for drying the body after a shower. Water (Hydrogen + Oxygen) covers our skin as we bathe and exiting a refreshing shower and drying the skin with a white towel interrupts the Hydrogen-Oxygen benefit.

Hand towels or wash cloths used will influence the electricity in the region of the mouth and hands. Pastels of green, pink, purple, blues or yellow are beneficial colors for hand towels and washcloths. Avoid gray, white, brown, black and navy towels.

Small white towels (bidet towels or baby wash cloths) should be used in conjunction with a bidet. Toilet paper irritates the bowels, kidneys and bladder and will eventually result in incontinence and alterations in female genitalia.

Not-So-Fun Fact: Paper towels interfere with the electrical charge inside cells when used for drying the hands. Use paper towels to wipe up spills or for cleaning purposes but they should not be used on wet skin.

CHAPTER XIII
TABLE LINENS

As discussed in the book Eating Yourself to Death published 2024, it is beneficial to not ignore the proper use of table linens. Cotton or linen are acceptable choices for tablecloths, placemats and napkins. Linen can propose problems when it comes to pressing the hard-set wrinkles but doing the best you can is far better than drycleaning with chemicals. Avoid synthetic materials that attract static electricity and interfere with the goal you want to achieve when eating. Synthetic is not something you would want to wipe your mouth with since teeth are a center point for electrical charge and synthetics can disrupt electrical activity. Eating is considered sacred in the energetic realm and should be treated properly to obtain the most benefit from the experience.

CHAPTER XIV
THE SIN IN SYNTHETICS

Cold feet? Someone in the generational line wore synthetic, rubber or other man-made materials on their feet. Swelling in feet and lower leg region can also be rooted in these comfortable, yet man-made and physical body altering shoes, such as tennis or athletic style shoes. I'm lacking research in this area, but I would say the gallons of water it must take to create those rubber or spongy shoes gives off a signal to the meridians in the feet telling them to accumulate water. Another component would be how the man-made products interfere with the Heavenly Gases, either by altering the measure thereof or eliminating one or more of them. Rubber bottom shoes can also contribute to static electricity, which damages the cells in the body. Personally, rubber sole shoes like what many tennis shoes have, make me physically tired. There is one popular brand of tennis shoe that causes me to feel as though I have two left feet! What is going on with that? Leather shoes accompanied by a leather sole will help avoid fluid issues that attract to the lower region of the legs and into the feet.

Ever wonder why the rough skin on your feet or elbows does not respond to creams and lotions? Materials not only influence the electrical highways in the body but can be a contributor to that rough skin that seems to not want to soften up!

Synthetic material influences the thighs and can be given an award for brewing up infection. Eczema and hives are easily irritated by synthetics.

This is a fact I did not know until I did a quick internet search on faux materials. Faux material has clear adhesives in it. If you are sensitive to adhesives, such as skin redness or rash where you've worn say, a Band-Aid, it may be wise to stay away from faux products.

Cancers and rheumatoid arthritis can result from synthetics, particularly Rayon and Nylon. Nylon also has an influence on the forebrain that can result in headaches and challenges with learning. If you have issues with headaches, it might be worth your time to eliminate all Nylon.

Synthetic materials in dark colors that host large flower prints can initiate an internal atmosphere that may be primed for strokes. Flowers and flower prints have an opening and closing or blooming type influence on cellular activity. Flower prints should only be worn during specific seasons and should be small flower prints. We don't want things expanding and contracting to a point blood does not properly pass through a blood vessel, vein or artery.

With the various ways synthetic materials are generated there are many signals that the body can receive that will manifest as a medical issue or irritating symptom. Everything from dental issues, to rocking the hormones out of balance can become your daily nightmare. Wearing any synthetic material in the color red will push irritations into the blood. I stick with this general rule, no red pajamas, robes, or undergarments. Red speaks directly to the blood and adding a signal from a chemically made material could be a disaster.

Flatulence can be a response to synthetic materials. When Heavenly Gases inside the body are disrupted, things become out of balance and gases collide. Synthetic materials cause a disturbance to the Heavenly Gases inside the body. One cannot help but envision the rumbling of tummies and bursts of fumes that come with the beautiful ballroom gowns that are on display at High School Prom or formal business events. A large percentage of those lovely gowns are made of a synthetic material.

Long ago, Satin was made from threads of the silkworm cocoon. Today, Satin can be made of Polyester, Rayon, Silk, Nylon, Viscose, Cotton or Wool. You will get what you pay for when it comes to Satin. Word of caution, Satin trim on a blanket is not advised, especially for babies.

Nylon contains plastics and will fade the color that is present inside a healthy, properly functioning body resulting in muscle disturbance. Leg muscle strength or coordination seems to go astray. No doubt this muscle influence could reside in the legs simply from the Nylons worn by many women to complete their business wardrobe. Nylons worn to the knee can result in the eventual need for a wheelchair. Situations where Edema is involved comes to mind. Nylon undergarments influence the movement of the energies in the spinal column and cause excess heat inside the body. With this activity, dirty fluid begins to take over that will eventually gravitate to the brain. A fact that should not be dismissed is the heart is a muscle and is likely influenced by Nylon as well. Mix Nylon with Spandex and the result is an interference with the ability to digest and process gluten. Synthetics inflict a hardship upon the sacral chakra (wheel) area resulting in a disruption in proper digestion and intestinal/bowel function.

Leaving the bum bare is an entirely different set of problems. A large bustline and/or bum are evidence that somewhere along the ancestral line someone did not properly cover their "privates" with undergarments made of natural fabric. The bum must be covered while sleeping or it will become large, although where we cross from perfect size to large is still unknown. More on this subject in Chapter XVI, Are You Sure You Want to Wear That?; subheading: Intimates. At this point, there seems to be no win for synthetics outside of the fact they rarely require ironing.

Polyester is an internal heat generator on its own and add black to the mix and it can eventually stir up trouble within the blood and runs a close race with Nylon in all the categories discussed in the previous paragraphs. It appears convenience (no ironing, washing machine safe, etc.) comes at a high price.

Rayon is made from cellulose or wood pulp and is known to contribute to cancer. In Home-Made Answers for Cancer and Life Altering Disease I added a photo of a tag found in the neckline of a Rayon button up, long sleeve shirt I had purchased at a boutique a few years prior. That tag stated: "Warning Cancer and Reproductive Harm." That alone describes the interference Rayon has on not only the color energy in the body but the electrical activity. Maybe my thoughts are a little different than most of the population when it comes to what clothing should be made of, but it would seem to be simple, common sense to mandate that clothing manufacturers are not allowed to make clothing out of products that are a potential health hazard. Shouldn't that rule be cemented into Basic Clothing Rules 101? At times it appears that common sense has flown out the window.

Viscose is a close cousin to Rayon, being made from wood pulp. Viscose and Flannels cause an increase in the dampness in the body and ultimately influences the blood. Flannel can be a blend of wool and cotton, or synthetics. The less expensive the article of clothing, the more likely it has synthetic material.

It is best to avoid Flannel pajamas and bedding. Flannel attracts static electricity and for the sake of having healthy cells, one should consider putting the Flannel to rest. Unless you find the specifics on what a particular Flannel item is made from, it would be best to refuse its comfort. Needing assistance in your elderly years may have a Flannel contributor as Flannel will influence the legs.

Tencel is reconstructed cellulose fiber that comes from Eucalyptus trees with a dash of chemicals added to the process of production.

Plastic is polymeric material that can be molded or shaped.

Fleece is mainly Polyester, polyethylene terephthalate (PET) synthetic.

The chemicals in Vinyl, a plastic material, influence urine production and elimination. Anyone with kidney or bladder issues should avoid exposure to Vinyl.

Personally, I stay away from all forms of synthetic material and recycled fabrics. Due to an unknown energy component that could remain in those popular recycled articles, I choose not to wear a garment that has been worn by others or doctored up by chemicals.

The dance leotard material and style is a recipe for obesity. The dance leotard look may be cute on a three year old, but they certainly are not healthy for them. I feel it best to avoid expanding on what impression the leotards make when worn by those supporting extras pounds, or years. What happened to dressing neatly and appropriately before going into public? This question reminds me how my grandmother would dress to go to the supermarket. If my grandmother was stepping out into public for most any reason, she would dress in a button up blouse with a pair of dress trousers and a suit jacket, no matter the season. Her shoes were black as well as her handbag, matching. Yes, this was several years ago as grandma was born in 1905 but maybe this selection of style for outings should be resurrected. There is something to be said for the fact she lived to be 107 and still had her memory.

When synthetic materials are worn on the body, they cover part or all of the spinal column. What are these toxic fibers doing to the spinal fluid that circulates around the brain? There is damage being done that either isn't being reported, being denied, or tests are not thorough enough to capture the disruption that would occur in spinal fluid. The last thing anyone wants is a situation where the spinal fluid becomes contaminated. This toxic fluid can lead to infections in the brain and decrease in cognitive function (Dementia or Alzheimer's symptoms). Many synthetic materials inflict a hardship upon the neurotransmitters in the brain and eventually results in obscure brain activity. Resting the head on a Polyfil pillow can result in sinus issues such as postnasal drip and stuffy or swollen sinus passages. Combine the Polyfil pillow with synthetic bed linens and the numerous electronics that seem to take up residence in bedrooms and one could end up with a disastrous situation.

Ever wonder what all those tiny fuzzy like fibers are you can see amongst the rays of sun? Could they be synthetic fiber fuzzies? Synthetic materials can sluff off tiny fibers that influence the lungs. Persistent cough? Synthetic materials may be to blame. This gives rise to the question as to whether synthetic materials are guilty of escalating lung infections or cancers.

Microfibers are notorious for harboring static electricity. Microfiber, the Polyester and Polyamide fibers that many use for cleaning the house seem quite popular. Has anyone considered what those microfibers could be doing to their health? These soft, dust attracting cleaning cloths conflict with the cells in the body creating a slow developing dangerous situation. The microfiber combination of chemicals is a disruption to the Heavenly Gases we need inside our body. Toxic chemicals colliding with the Heavenly Gases or even with one another results in flatulence or painful intestines. Would there be dangerous situations when it came to the hands that hold these cloths, or could this damaging activity happen within the entire cellular system? This is a disturbing thought if the damage can initiate in the hands and travel up the arms. Hands, like the mouth, can be an inlet for things to enter the body. The answer to this will have to wait until accurate studies are performed.

Corduroy blocks the ability for the body to receive adequate chemical elements and interferes with the activity of the Heavenly Gases. If you are challenged with your energy level, the culprit may be Flannel or Corduroy. Both of these fabrics influence the muscles in the legs, causing the muscles to become rock hard over time, or make its debut later along the generational line. Stylish as they may be for the days of crisp Autumn and cold Winter, you will need the full function of your legs, not only now but later in life!

Denim is made of cotton but what is it about denim that gives rise for a concern? Thick, tightly woven fibers that create the barrier described for corduroy. Fine, soft or luxurious fibers are easier for the body to accommodate. Dark blue is not an easy color for the body to accommodate, and to top it off, fabric such as denim could very likely be one of many culprits that cause wrinkles in the skin!

Athletic clothing wins the award in the synthetic category. The wicking, cooling, no sweat and whatever else they have come up with is an invitation for altered cells and eventually damage to the DNA helix. Polyester, Nylon, Spandex should all be eliminated if you desire to be healthy or remain healthy. As you heat up your body and the pores in the skin open, your body absorbs a percentage of the petroleum, plastic and other chemicals that make up those stretchy, skintight coverings on your body. As stated before, Nylon is plastic, and if you have read Home-Made Answers for Cancer and Life Altering Disease by Harvest of Healing, published 2024, you have been enlightened as to what plastics can do to the brain. Plastic can become stuck in the brain and interfere with electrical activity necessary for proper body function.

Fun Fact: All those hours of exercise while dressed in the matching stretchy synthetics that eventually spell out obesity, just might be the reason for your need to be at the gym.

Those who spent their childhood years in the 1970s or 1980s probably had a handmade crocheted blanket from their aunt or grandma. The gesture was nice and heartfelt but the Acrylic yarn those blankets are made of can cause havoc on the interior of the body. The result can be issues with the legs, right where we would put the blanket to keep us warm while watching television or nursing a cold. Acrylic: a transparent thermoplastic material. This could explain the arthritic hands Grandma would take Aspirin for.

Quilts are most often filled with Polyester batting, the less expensive way to fill the quilt blocks. Couple that with the mix-match of patterns and colors normally found in quilts and the body is bombarded with numerous signals. Again, the legs will pay a price for these combinations.

Potential Symptoms

Enzymes in the body are influenced by these manmade materials discussed above. With the numerous digestive issues I experienced in the past, this fact makes me wonder if the denim jeans or synthetic shirts I wore contributed to my lack of digestive enzymes. Gluten intolerance plagued me for about ten years, during the years I wore clothing made of the troublesome materials. Over time and through the gradual process of altering my wardrobe and gluten is no longer an issue for me. Coincidence? I don't know that I have the answer, but I doubt it.

Sheer, see-through materials or fabrics brew up the genetics for rheumatoid arthritis and genetic homosexual tendencies. To avoid an overload of chemical elements to the body, see-through materials and fabrics need to be avoided.

*Gluten intolerance and digestive disruptions.

*Fluid retention; clogged and polluted lymph

*Internal heat regulation disturbance.

*Issues with bowel leakage.

*Sore or weak leg muscles.

Not-So-Fun Fact: It can take up to five years to clean damaging influences from synthetic materials out of the body and the cleanup phase can be challenging at times.

CHAPTER XV

DOING THE LAUNDRY

There are a few unknown, or at least not commonly known, rules when it comes to laundry. Laundry not only applies to the clothing you wear but also to the interior of the body. Medical issues will result for not wearing properly washed clothing.

Plant based detergents are a more gentle option for clothing and not as likely to fade the color. Enzymes added to a detergent or applied through a means for stain removal will gradually fade the color of your clothing so go easy on the detergent if it contains enzymes. Appropriately stain treat items when necessary.

All detergents and fabric care products should be kept in a drawer or cabinet and not left out on the laundry countertop or on top of the machines. Fragrance free laundry detergent is the safest route and sheep wool tennis balls spritzed with your favorite essential oil(s) do well for the dryer.

Wash undergarments, including socks, slips, t-shirts, bras and panties after each time they are worn. Wash intimate apparel separately from other garments with a gentle detergent. When possible, line dry your intimate apparel. Men's business dress shirts should also be laundered after each use. Towels should be washed separately from other laundry.

Clothing should not be left on the floor. Leaving laundry on the floor will result in dirty fluid collecting in the head. All dirty laundry should be immediately placed in a laundry basket or hamper. Laundry baskets are designed to hold dirty laundry or transport laundry. It is beneficial to not use the same laundry basket for collection of dirty clothing and for transporting clothing once clean from the dryer to a surface for folding or placement in a dresser or closet. Use separate baskets for these chores. When storing your clothing, it should be grouped by color. A closet or dressing room with sufficient lighting is important when it comes to matching the colors of your wardrobe selection for the day. Closet doors are preferred, and when possible, walk-in closets can become a usable dressing room.

All laundry should be folded and put away directly after drying. Fold your clothes neatly with the seams matched. I find this is challenging when folding fitted bed sheets. The elastic edging throws a wrench in this matching the seams theory. As my Grandmother told me in this situation, "do the best you can."

The dining table it not to be used as your holding or folding station for laundry. Dining tables must be kept orderly and piling clothing on the dining table could influence not only your mealtime but also signals emanating from the table to the clothing.

Ironing creates a harmful influence on the level of Heavenly Gases within the body. Try to keep ironing sessions brief and be sure to iron all clothing you plan to wear on Sabbath rest day the day prior. Heat is to be avoided on days of Sabbath.

Avoid dry-cleaning as much as possible. Wool should be spot cleaned. Any chemical introduced to a natural fabric will damage the natural vibrancy in the fabric.

No stain guard or scotch guard type substances, and certainly no flame-retardant chemicals in your clothing.

Fun Fact: For the best results during your sleeping hours, pajamas and bedding should match in shade or color.

ACT III
STYLE

CHAPTER XVI
ARE YOU SURE YOU WANT TO WEAR THAT?

Many of the chemical elements that are present in the atmosphere and necessary for the health of our body are sensitive to not only our activities but the way we are dressed. Some of the information shared about the influence clothing can or will produce may seem illogical until consideration regarding the internal gases (Heavenly Gases) is included in the equation presented.

Designed for Male? Or Female?

I will address the tender subject of cross-dressing first. Cross-dressing will afflict the Soul. What do I mean by this? Male and female electrical activity is different on levels that have yet to be discovered by a means of scientific study. When clothing designed for the female is placed upon a male, or vice versa, the messages sent to the body through the clothing can disrupt functions naturally assigned to the blood. Remember, the goal is to keep the proper amount of light within the plasma and cells, that light is the foundational source for the Soul. When that light (Star Dust) is damaged, the body struggles and will eventually begin to develop signs of disability or disease. When a person exhibits a desire to clothe their body in fashion designed for the opposite sex, there is a disruption in the blood that is initiating such a desire. I do not personally know anyone in the category, but it is challenging for me to accept that such an activity is solely due to a controllable decision. Something within the intricate details of blood codes and DNA is at play that apparently becomes soothed by clothing oneself beyond the gender specifics. We are all electrical beings and on a general scale an electrical hiccup is at play in these situations.

Fun Fact: The vessel in which each Soul resides must be protected from outside electrical interfering invaders.

Mixed Messages

Certain elements of the fashion industry have led people astray simply through colors and patterns. Those misconceptions that are spread across advertisement airwaves and draped upon those who decorate magazine covers and parade past television screens can be impressive, but they are not always beneficial. And what might benefit a popular figure may not benefit you, dashing in appearance or not. Caution should be applied when deciding when or what fashion to follow.

Specific meridian juncture points can leak the electricity from the body when they are not properly protected. Covering the body properly, with clothes or fabrics that help neutralize harmful frequencies or produce/protect the beneficial frequencies is required to keep from losing the power within. When unwarranted episodes of feeling tired have been experienced, the cause may be contributed to the lack of proper covering or simply wearing a color that does not correspond with what is needed for that day's events.

While black can prevent the blood from singing its healthy tune, white prevents the internal cleansing process from functioning properly. Excess weight can be a sign of a malfunction in the internal cleansing process.

Stripes, polka dots, paisley or plaid, leopard, cheetah or zebra patterns should all take a bow for disrupting internal electrical activity. Solid colors are best when it comes to energy.

Styles from the 1970s need to be tossed out. Some of the fashion designs from the 1970s are as bad to the inner body as the numerous pop-ups on a computer screen is to the brain – too much information all at once! The tie-dye, fringes, large flower prints, hip-hugger jeans, uneven hemlines and bell-bottom pants are just a few fashion statements that need replaced. There is nothing orderly in tie-dye and the movement of fringes can be an invitation for instability in the area of the body the fringe is worn. As mentioned previously, large flower prints will cause the body to bloom! Most often in places you would prefer it not.

Strapless or spaghetti strap styles place a burden on the region of the shoulders, this would include bra straps. A proper bra style consists of tank top style shoulders.

Your clothing should be in good condition. Thin or overly worn fabric, holes in fabric, hemlines that are uneven or not properly secured will throw off the movement of the energy within the body and can result in loss of energy you need for your day. Caffeine should not be the go-to for an energy boost. Keeping clothing in its proper and beneficial condition is where energy focus should start. Thin or worn fabric will influence the left side of the body and the knees. Avoid clothing with cut-outs, partial back or sides cut out, etc. These are access points for harmful disruptions to begin and they do have an influence on the blood.

Not-So-Fun Fact: Paisley prints remove the color that is produced by the Heavenly Gases inside the body.

Blouses and Shirts

Shirts are to rest below the waistline, otherwise shirts that rest at the waist cause excess weight to collect in the form of no defined waist or a belly pooch. Keep the belly in good condition by tucking in your shirt. Exposure of the belly will interrupt the ability to absorb the Heavenly Gases and puts a strain on the digestive system. An exposed shirt hemline causes energy to accumulate or be intensified at that line of transition from shirt to trouser or skirt. Tucking the shirttail into the pants or skirt creates a smooth transition from top to bottom.

Round edge collars create a softening on the edge of the collar and are preferred over those that come to a point.

Sleeveless clothing gives opportunity for too much air to come in contact with the arms. Too much air results in fluid accumulation and evaporation of the Heavenly Gases. When the fluid volume goes up, the gases decline. Sleeveless clothing should be worn with a jacket or high-quality sweater. The arms should not be exposed to cool air and to avoid phantom pain in the shoulders, appropriate protection from the outdoor elements is required.

The history of wearing a blouse or top with a boat neckline, V-neck or other neckline that exposes the clavicle results in an elevated clavicle and sensitivity in the area. A button down, long or three-quarter length sleeve shirt with a pocket placed on the upper left (over the heart) creates adequate energy movement along the meridian highways.

Ribbon or flowing pieces over the shoulder reflect instability and will negatively influence the region of the shoulders. Shoulders are a reflection of stability or endurance, an ability to carry the load. When a person cannot carry their load the door for emotional breakdown occurs. It is advised to avoid dangling or flowing articles at the shoulders. It is considered taboo (outside of the boundaries of social custom) to wear blouses or casual tops that are made of a see-through fabric. Lace across the chest interferes with the function of the lymph.

Ruffles on clothing will cause a burden that will advance into pain in the region the ruffles are worn. If you have a blouse that has ruffles over the shoulders and down the front it can result in shoulder injury or pain. Ruffles conflict with the orchestration of healthy cells. By healthy I mean cells that contain the appropriate balance of Heavenly Gases to produce light. Simply stated, the body will not respond well to flowing or dangling pieces of material or fabric.

Most sport team jerseys are made of synthetics and adding a name, logo or number adds complications to an already potentially damaging situation just with the synthetic material. Lettering will act as a burden to the region of the chest. This includes all sports related jerseys or uniforms.

The lymph system can become impacted by these jerseys, resulting in an inability to move dirty fluid out of the body. That dirty fluid eventually becomes part of the cells. For the sake of a healthy lymph system, avoid clothing with logos or lettering.

Dresses and Skirts

Long formals are often worn at night, the time when the influence of the moon is most evident. Combine dresses that sweep the ground, particularly when going up or down steps, with the moon energies at play and internal chaos can begin. We must remember, the earth itself contains electrical charges and dragging clothing along the ground or the floor could easily attract an electrical charge that interferes with the meridians in the body. As the Biblical story of the woman with the issue of blood reflects, the hemline is representative of the condition of the blood.

Button down the front dresses display a sense of openness to anything that comes along. The Conception Vessel meridian runs from the indent just above the chin down the center of the body. This meridian is influenced by dresses that have no secure seam along the front line of the body.

Wearing a skirt with a blouse is considered to be a sign of lower status than wearing a dress. A skirt should be accompanied by a matching jacket and are intended for professional business. Business suits with a matching jacket should be worn in all professional buildings.

Pleated skirts initiate creases and folds in the body. How this plays out in the interior of the body will be an interesting discovery. Can the DNA strands become creased? A quick google search states they can, giving rise to mutations. Exterior creases would equate to wrinkled or baggy skin. Pleats also contribute to having a flat bum (an act of folding down/over)!

A double message is sent to the body when a skort is worn. Too many messages (electrical in nature) coming at the body at any given time can brew up an infection.

Antique white is most beneficial when selecting a wedding gown, not white. White, as previously stated, should be set aside. Wedding gown color and styles need a makeover, eliminating the floor sweeping hemlines and Pearls sprinkled about, not to mention the synthetic material. Gowns in pastel colors would be uplifting for any bride on her busy day. Remember, the idea is to attract the chemical elements from the atmosphere through the color of our clothing to keep the body healthy.

Short dresses, long shirt style dresses or short skirts create a hinderance. Legs are sensitive to cold and sun, and meridians that run along the legs will respond accordingly. Ladies, exposing your legs to the sun will result in hair growth on the legs, including the thighs. Any skirt that exposes the thighs will result in energy disruptions that produce large thighs. I think everyone would agree they had an ancestor that wore shorts at one time because every woman I've known has some measure of hair on their legs. We may as well wrap wearing shorts in with the short skirt description. Excess hair growth, large thighs... what else could develop from exposing the legs? Spider veins. To think that prior to the introduction of shorts women, and possibly men, had naturally clean, smooth, hairless legs. Knee to mid-shin length dresses and skirts protect the kidney, stomach and spleen meridians.

HEMLINE should be up off the ground to prevent dragging. Accumulation of soil on the hemline will result in toxic or sluggish lymph and dirty plasma. Hemlines should have a depth of 2 ¼ ." No fringe and no uneven or unfinished hemline. Fringe creates genetics for being overweight and convey a message of being unstable. Fringe may add a little flair to the outfit, but it also impacts the cells in the body. DNA strands and meridians are influenced by every cut, fold, crease, wrinkle, and so forth. A crippling influence occurs when hemlines are not proper and up off the floor. This crippling lies beneath the surface in the blood. When meridians are hindered, the resulting impact will gravitate to the blood. Protect the electric highways that run through the knees by keeping the hemline of a dress or skirt to just below the knees or mid-shin. The highways near this area govern the spleen and stomach. The crown of the head and color of the eyes are both influenced by the hemline.

Trousers (slacks or pants) and Shorts

Not-So-Fun Facts: Trousers that button at the waist will cause a ballooning of the belly region. Pants made of a flower print material or fabric can result in bowel leakage.

Drawstrings at the waist cause the waist to reduce in size and the belly to become large. If you do not desire a pooching belly, it would be best to retire the sweatpants.

No leggings or skinny jeans. Leggings not only influence the shape and function of the legs but are not appropriate attire if you desire to keep your legs lean and in shape. Spandex, Nylon, Elastane, Polyester, all work against what you are attempting to accomplish at the gym when it comes to toning the legs.

Ladies, wearing trousers should be limited to outdoor events or working in the yard. Otherwise, dresses or skirts are in order. Yes, I wear trousers on those stay-at-home days when it is extremely cold weather, but I am sure to wear a good quality (cotton, wool or cashmere) dress style trouser with casual heels. Windy Spring and Summer days also warrant a day for a pair of dress trousers. I have a linen suit that works well on windy days that require exposure to the outdoors. Last thing we need is a Marilyn Monroe event where our skirt flies up in our face!

Trousers should fit comfortably, nothing tight or snug fitting and certainly not baggy. Remember the days of the extra-large trousers many young teens, mostly male, supported? When it is a requirement to hold onto the waist of your pants to avoid them falling to your ankles, they are too large! Baggy trousers cause heat issues in the lower region of the body. Sometimes you must wonder where the idea of a specific fashion came from. Who would have thought trousers, usually jeans, would be appropriate attire or even comfortable when worn two or three sizes too large! Wearing the PROPER trousers and underwear will increase the ability for the Heavenly Gases to do their job within the body. It is a product of wisdom to make sure clothing fits properly.

Men, a zipper fly on trousers or shorts will cause an unhealthy electrical disruption to the region of the groin/penis. Appropriate attire for men is trousers, no shorts.

Men and women's business suits should be a matching set. No mix-matching fabrics of the clothing you are wearing.

There should be no rear pockets on your trousers. This is a challenging issue to avoid. I have purchased dress trousers that have pockets on the buttocks, but I leave the stitching to where the pockets remain closed. The location of trouser pockets on the derriere can result in fluid issues in the body. Pockets will also trigger meridian points, and any fluid response is likely connected to the bladder meridian activity. No lace or decorative beads, or sequence on rear pockets.

You will invite electrical disruption if you wear garments with holes, tears or damaged seams. Thin fabric in the area of the seat of your trousers will influence the left side of the body.

My dad was born in the late 1920s, and he told how at the end of the day if your trousers had any form of tear or hole from the activities of the day, they (he and his siblings) were to put the damaged trousers in a basket beside their mother's chair prior to turning in for the night. My Grandmother would mend the pants after her children went to bed being sure to have all trousers patched and ready for the following day. Dad never was fond of jeans with holes or tears. An ancient secret may have been attached to his dislike for overly worn or torn clothing.

Ladies, a blood temperature difference exists between days that you wear trousers versus days that you wear a dress. I have no calculations for this but what comes to mind is when young ladies may be challenged at becoming pregnant. Temperature regulation can be a key when attempting to conceive.

Dressing up each day keeps the day moving along smoothly with protection from harmful disruptions in the energy flow of the day. Sweatpants and denim (jeans) are disruptive for Spiritual Royal blood. Spiritual laws exist regarding the clothing worn and consequences will arise somewhere down the line when guidelines are abandoned. Dressing in distaste or sloppy in appearance will come with a price to be paid in the health of the plasma that ushers in urinary and bowel disruptions. There are likely many more consequences that have yet to be revealed. Some things just do not play well with nature and its forces and sloppy clothing is no exception.

Fabric Patterns

Plaid will cause the region of the chest to be exposed to potential harm. The heart comes to mind given the electrical activity involved. The crossing lines of plaid interfere with the inner cellular activity that is to take place during the season of Christmas, or Winter Solstice. The region of the right breast can also become influenced. The negative/positive charge in the blood is influenced by plaid and explosive issues can result from these horizontal and vertical lines. It is as though the intersection of these lines cause an electrical short in the activity of the meridians.

Blue with green plaid is particularly harmful, again considering the influence on the fluid (blue) in the body. Argyle influences the opening and closing movement of the cells. In general, I believe printed fabrics are a challenge for the function of the meridians. Fabric with print is not for occasions related to business or formal celebrations.

Stripes can create another set of challenging issues for the body. When meridians in the body flow in an up and down pattern the display of pin stripe or broad stripe can influence those meridian electrical highways. Horizontal stripes create opportunities for the body to expand its circumference. It is becoming quite clear that what we wear on the exterior produces guidance to the activity on the interior. Where does this put the leopard prints and tiger or zebra stripes? Temperature regulation in the body comes to mind, the hindered ability to reduce heat or properly maintain a healthy internal body temperature. The question of whether there is an affliction on the thyroid function with these prints comes to mind.

Floral prints have much to say and have their own unique set of rules. Such as, to wear a floral print is indicative of spiritual growth or fertility that is current and still to come. The education regarding the governing forces of the Universe or, the natural order and orchestration of all things on earth, has nothing to do with educating oneself in or participation in any organized religious practice. This education comes with experience. God is the professor, and you are the student. Graduation from this education comes after the age of 50 years old. Whether you pass the courses or not is another topic. Given this, floral prints are to be worn by females within the age range of 0-50. I specify females because the role of a female on earth is far different from that of a male, some in obvious ways and some in mysterious ways. Expansion on this subject will need to come with a future publication. After age 50, solid color clothing is to be worn as an indication of maturity in what I will call the Universe Government Education program. The growth and maturing process is to be well established by age 50. Floral prints worn after age 50 initiates the continuation of maturing or, aging. Please keep in mind floral prints are for Spring and Autumn only.

Considering the information shared in the previous paragraph, Hawaiian print clothing causes confusion to the processes the body goes through in a day. Flowers are a powerful symbol and adding multiple colors to the display can result in genetics for homosexuality. How? I do not hold any scientific explanation but in the language of the Universe an open bloom represents the female genitalia, the collection point for seeds of the flower that eventually produce fruit (or more flowers). Flower prints can initiate the movement of opening and closing of the cells. Flower prints are a contributor to the continuation of the aging process beyond its borders.

Tiny flower prints are the best for not interfering with the size or shape of the body. Large flower prints have a reputation of causing the body to expand out, usually in areas you would rather not have the expansion. A good example is large floral prints on a stuffed chair or sofa results in genetics for a large derriere. Now you can blame great-grandma and all her floral furniture for that oversized bum you have been attempting to downsize. Many common Spring prints support a floral display. A floral print larger than say a small coin, can throw off the season cycle in the body. There is a specific time the body has for opening and closing of cells and when the body is thrown off of the natural cycle, the cells can end up full of debris that can result in health complications. As fun and colorful as they are, simply put, there is too much information coming at the body through busy floral and Hawaiian prints. The noble metal gold will help calm the influence of flower prints.

The cell opening cycle reminds me of Lyme's Disease, which was unidentified or denied by the medical industry for many years. From personal experience, Lyme's Disease does exist, and it can be very debilitating. How it is contracted could be related to this untimely opening and closing of the cells described above. If a large floral pattern is worn, particularly in the incorrect season, that floral print can trigger the body to have untimely cell activity, opening an opportunity for unwanted invaders into the cells.

No Easter prints, such as bunnies and eggs. Instead, opt for dresses in Spring colors. Easter prints contribute to the development of rheumatoid arthritis and can damage chromosomes, depending on the design and color.

Native American patterns and prints will cause you to look older than you are. Native American tribal attire has had a negative impact on the blood by driving it away from the origins of the DNA. Anytime the DNA is altered it burdens the body's natural processes. Feathers from large birds emit gases that influence the balance of the Heavenly Gases within the human body. This alteration causes changes in the DNA that spill out in skin pigmentation, hair texture and color and eye color changes.

Fun Fact: Depending upon the type and style of eyelet in the collar area of a shirt, it can represent an upper-class status.

Intimates

Production and movement of energies will be influenced by type, style and color or pattern of your undergarments and evening wear. Nightgowns and pajamas are bodily covering designed to be worn in the privacy of your home or in bed chambers, not to the shopping mall, on an aircraft or in any other public place.

If you have a habit of sleeping in the nude, you may want to reconsider. Sleeping in the nude causes the cosmic energies to collect in and influence areas you may not want altered by their power. Thighs and hips are common collectors for any excess nighttime energy, that will result in enlargement in size in these areas. If you wish to avoid altering or expanding your physical shape and size, it might be wise to limit nudity to the confines of a shower. Bowel activity also takes a hit when one sleeps in the nude. Best advice, cover yourself appropriately. Nudity while sleeping hinders the energies that recharge the body during your peaceful slumber. Nightgowns are preferred and should not have decoration, lettering, lace, etc. across the region of the chest.

No sleeping in suggestive, frilly, lacy or racy nightwear. The electrical messages from the fancy undergarments and nighties can build into pancreatic cancer if given the chance. The intimate wear classified as "risky" may be just that. You may attract unwanted issues that result in an affliction to the body. Could this lacey, racy and risky style be a contributor to breast or reproductive organ cancers? Future testing may bring us the answers to such questions. Lacey undergarments have produced promiscuous personalities and fluid accumulation along the spine that can eventually spell out incontinence. An important sidenote to interject here, any intimate activity should not involve being fully naked. For electrical function and energy's sake, keep the upper portion of your body covered.

100% Cotton attire for sleeping that supports the color of your bed linens is most beneficial. No white, gray, navy or black for sleeping, or anytime for that matter. Proper length of a nightgown is to the shin to avoid the thighs from becoming thick. Cotton helps erase any genetic imprints you inherited from a grandma (or other ancestor) who wore synthetic or lacey intimate apparel. Linen is a good helpmate in restoring health to the genetics because it speaks to the chromosomes. Linen and pure silk are permissible for sleeping. Although, make note that ankle length silk nighties influence the thighs by making them large. Bamboo is overly processed and is not a beneficial material. Blue lace nightgowns cause fluid to migrate to the brain.

Bras and panties should match, and bras should be made in a tank style to avoid phantom shoulder issues. A camisole should not have lace or spaghetti straps. Lace across the chest influences the function of the lymph and those spaghetti straps, just like on bras, will result in shoulder discomfort or pain.

Dress with multiple layers that include white t-shirts, proper underwear, a slip or petticoat. Layers provide more energy protection for the body. Being seen in just your underwear will have energetic consequences that cause the belly to balloon. Bra liners in clothing will result in black hair on the area of the clavicle (this would hopefully be restricted to male descendants). Nude or beige colored bras will result in descendants with small breasts.

Ladies, it is wise to own numerous panties and bras. Clean underwear in the morning and clean underwear in the evening is the order. You never know when you may need to be away from home for an extended time, or when there is lack in time to do the laundry, not to mention an unpredictable necessity for changing the undergarments might arise, so be properly prepared. Underwear should rest below the waistline during Waning Moon. During Waxing Moon if you prefer your underwear to ride at the waist, it is permitted. Underwear worn to the waist during Waning Moon is a recipe for a large belly. Stay clear of red, lace detail or see-through undergarments.

Universal colors of white, ivory or natural, and pastel pink all work well. Pastel colored underwear that matches the nightgown has a benefit during Waning Moon phase. Black lace clothing, specifically bras and panties contribute to Alzheimer's and Dementia, an obvious brain disruption happening with this combo. Petticoats or bloomers made of 100% cotton; bras should be made of cotton, no Nylon or lace. White or ivory, sleeveless cotton undershirts should be worn under all shirts or tops that are light weight. No colored undershirts. White, natural or the recent color "cloud" are all acceptable. Thong and bikini style underwear result in a large bum! Finally, it is taboo to wear negligee style undergarments under a dress.

I Timothy 2:9: *Also, the women are to dress themselves in modest clothing, with decency and good sense, not with elaborate hairstyles gold, pearls, or expensive apparel but with good works, as is proper for women who profess to worship God.* (CSB)

Genesis 9:20-25: *Noah, a man of the soil, was the first to plant a vineyard. He drank some of the wine, became drunk, and uncovered himself inside his tent. Ham, the father of Canaan, saw his father naked and told his two brothers outside. Then Shem and Japheth took a cloak and placed it over both their shoulders, and walking backward, they covered their father's nakedness. Their faces were turned away, and they did not see their father naked. When Noah awoke from his drinking and learned what his youngest son had done to him, he said: Canaan will be cursed. He will be the lowest of slaves to his brothers.* (HCS) (More detail is given on this story in the Book of Jubilees)

Ladies, please remember it is not proper etiquette to be seen in your undergarments by anyone. The magnetic pull between one person and another can deplete beneficial energy or present a situation where you receive a dose of harmful influence from the party who sees you. An intoxicating situation in the blood occurs as reflected in the verses in Genesis above. Exposing oneself to the view of another while only supporting your nightgown or undergarments will result in a large waist and protruding belly. The body does not respond well to being naked so please, for the sake of the womanly figure, remain discrete. This will keep the legs looking good as well. School and sports event locker rooms need a makeover to prevent dressing or undressing in front of others. Modesty pays off! Dress in a designated dressing room or closet area. Casting the eyes on a naked body will result in issues with your vision.

Formal Appearances for Female: After proper washing of the face and hands, approach the dinner table in proper dress: White t-shirt and undergarments to accompany your long dress or skirt. There should be two layers of fabric under the skirt. This double layer can be in the form of a slip under the skirt or dress. Formal dresses, nothing that drags the ground, made of a solid color are the most beneficial for your evening. Wear short, evening style fur coats for evening dinner engagements. Button closures are preferred over zippers and a blouse should button in front. Do your best to stay away from zippers or snaps. Yes, this can be a challenge. If you need to attend a business meeting or formal event that does not include dinner, a suit with a skirt and jacket serves business meetings well.

Male: Men, your selections are very basic. 100% cotton boxers and white undershirts, no colored undershirts for business attire. Boxers or briefs should be a solid color. Business suit with white or pastel business shirts and a colorful tie. Neckties should not have a loose back tail. The back tail of the tie should be tucked and/or secured to the front of the tie with a bar or clip. Neckties are required for those in professional work or official positions.

Coats or Jackets: Ladies, a suit jacket is to be worn with all skirts when inside a professional building or attending a professional function. Cardigan sweaters are good as a general additional layer.

Fun Fact: Any pantyhose style stockings, preferably silk, must be accompanied by a pair of underwear.

The Little Ones

What about babies and children and their undergarments? No white onesies for babies and please do not dress a baby in clothes with stains on the collar. Babies and children are more fragile when it comes to the intricate electrical system in their bodies. Adding a lightweight beanie or cap helps protect the crown of their head. Ruffles on the bottom of bloomers, as cute as they are on those baby girls, can result in a large derriere. Ruffles signal the body to expand in the area the ruffles are being worn. Ruffles on the bottoms of baby bloomers or little girls' panties can also result in bowel issues or leaking feces.

Who doesn't enjoy a little play time on the floor with the infant or toddler? Ladies, remember to sit in an appropriate manner while on the floor, keeping the legs together and tucked in a secure fashion. No sitting cross-legged, especially when on the floor in a dress!

ACT IV
ACCESSORIES AND SHOES

CHAPTER XVII

FROM HEAD TO TOE

Hats, Belts and Sashes

Cowboy hats interfere with body temperature regulation and fluid. Hats made of beaver hide will initiate fluid issues and hats made of rabbit skin will influence the function and health of the ears. The shape of the top of the cowboy hat where it has a dip in the region of the crown of the head may be a contributor to the issues described. Hats made with synthetic fibers influence the bowels and cause disruption in healthy brain signals. Avoiding hats with a metal button in the top may be wise, like those on baseball caps. The metal crown button can receive an electrical charge that will influence the crown chakra at the top of the head. Hats made of paper give me a headache. A fluid issue may be to blame since paper is made with a fair amount of water giving rise to a water signal penetrating to the brain.

Unacceptable materials can also be found in hats or head coverings. All head coverings and hats should be made of an all-natural fabric, such as wool, or even straw for those hot summer days. When possible, large brim hats are most beneficial.

Add lettering or symbols to a hat and the chances of fluid accumulation in the brain increases. The Universe reads letters as symbols and there can be more of a message in that baseball cap with your favorite team logo on it than we know at this point. These hats that many heads support will eventually influence the vision. Yes, it is important to protect the head but do so with quality wool and no symbols or lettering on the hat.

James 1:11: *For the sun rises with its scorching heat and dries up the grass* (flesh); *its flower* (prime of life; youthfulness) *falls off, and its beautiful appearance is destroyed. In the same way, the rich man will wither away while pursuing his activities.* (HCS) (descriptions added)

This verse in James lays it out, wear a hat if you wish to keep the flesh youthful. Ladies, with an expanded study which this publication does not support, the flower is symbolic of the female reproductive organs. A "bud" references prior to fertility and to be in a state of "bloom" is a reference to the years of fertility. Once the monthly cycles have discontinued, the blooming flower has graduated to fruit. Once the age of a woman reaches "fruit" it is best to limit wearing floral prints. If floral prints are worn, be sure to wear the small floral prints in the seasons of Spring and Autumn only. Side note: avoid exposure to the sun that causes the physical "withering" of youthfulness. The health of the crown of the head (where the sun would hit) connects to the health of the female genitalia. How? The Governing Vessel meridian runs from the top of the head through the body to the region of the tailbone. What goes on at one end of this meridian highway will influence the opposite end! This "from one end to the other" passage gives rise to the thought of the many chemicals applied to the hair. That new color or fanciful hairdo with many gels and sprays may not be as important as once thought! Is such activity introducing the possibility of reproductive cancers, fibroids, menstrual cramping, etc.?

To properly finish off your wardrobe, define the waistline with a leather belt, or matching sash tied in a bow. There should be no dangling extensions proceeding from the waist, such as fringe or tassels.

Handbags, Purses, Backpacks, Fanny Packs

What is carried in your purse or handbag influences the body. Tampons and pads with adhesives contribute to genetics for blonde hair. Anything carried in the blue handbag will influence the fluids in the body. The body has a rhythmic clock that controls the ebbs and flows of the fluid in the body and this natural rhythm can become altered by exposing the body to consistent shades of blue. Wearing a blue uniform daily will produce some form of influence to the fluids in the body. Black handbags are the color of choice. Handbags or purses should be made with double handles and some form of snap or clasp for securing a closure. Handbags with shoulder straps result in genetics for phantom shoulder pain. Shoes and handbags are to match. Store your handbag or purse with your wallet in a drawer.

Fun Fact: Large handbags result in large people.

Footwear

Shoes should be made of leather, including the soles. No flip-flops or sandals with a strap or divider at the big toe where meridians for the liver function are located. Sandals should have an enclosed toe to protect the meridians that are present in the toes and feet, which include stomach, spleen and liver meridians. Black baby doll or Mary Jane style shoes with a buckle and white socks are most beneficial and aid the body in electrical balance. Red shoes are worn with red clothing only.

Navy shoes should be worn with clothing in shades of mauve. Pink shoes appear dirty when worn with mauve. Brown, like moccasins, will cause the skin pigmentation to become darkened so limiting brown shoes is best to avoid an overload on the cleansing process the body will go through to rid the body of the signal from the color brown. Avoid wearing white shoes that will appear gray when worn with white dresses or trousers.

Wouldn't it be odd to discover whether the teeth can be influenced by the type of shoe a person wears? Has your dentist ever asked what your shoes are made of, or what style they are? Shoes have a big influence on the body and protect us from some of the electrical activity that takes place within and under the soil. Teeth hold electrical junctures and with the number of teeth in a mouth and the electrical activity we encounter throughout a day, it would be no surprise if shoes, and clothing, had an influence on the health of our choppers!

Women should wear two-inch heels. Try to find heels that make minimal noise as you walk. If your home or office has floors covered with carpeting this will not be an issue but if you spend most of your time walking across hardwood or tile, the noise made by the heel of the shoes as it meets the flooring will record in the cells of the body. Wearing heels not only tones the legs but produces a signal in the body to keep things elevated. This is a fancy way of avoiding a few bags and sags that seem to come with the aging process. Shoes have a big influence on the heart and heart meridians, and condition of the cheeks, both sets! Wearing heels speaks a message to the blood to keep the facial cheeks full. This same message is sent to the "cheeks" on your backside that you sit on. The bum will droop when wearing heels is eliminated, when you, or your ancestors, sit in folding chairs, and when you stand at the kitchen sink with your hands in running water while doing the dishes in bare feet, or wearing those flats, or tennis shoes you can't seem to give up. On occasions that require you to stand or walk for extended periods of time wear a half size larger shoe than you normally would.

No leather or synthetic material tennis shoes. Leather tennis shoes, rubber sole shoes, yard shoes and not matching shoes to an outfit all contribute to obesity in women. Leather, when accompanied by a rubber sole will influence the temperature regulation of the body causing it to be too cool. Canvas tennis shoes should be made of natural fabric and natural rubber, no recycled plastic, although a canvas tennis shoe picks up every energy it encounters just like a sock would pick up everything it walks upon. Canvas shoes have a reputation for influencing the lymph activity that may come from the sole of the shoe rather than the actual canvas fabric. Wearing any form of canvas shoe drains my energy and the following day I experience being very tired. Things as simple as the print inside shoes can throw off the energy.

Flipflops influence the face causing the skin to display a reddish tone and elevated heat in the body.

Slip-on style shoes produce a signal for instability on the feet that in later years can result in a need for a cane or other form of assistance while walking. Covering the feet in socks only (no shoes) will result in the same, along with producing over-exaggerated footsteps.

Avoid satin on the feet. Satin produces static electricity and mixing that with the electrical charge that is present in the carpeting or soil of the earth can create an overload of electrical activity through the feet and legs. This could be a contributor to the legs or feet being tired or experiencing discomforts.

Years of wearing ankle high boots will result in a roller-skating type movement, not being stable on your feet. Leather boots that go above the knee contribute to the alteration of Heavenly Gases within the body. This is a subject that will need more insight as time moves along.

All the ladies say hooray in response to the instruction to replace shoes at least every five years. Shoes begin to smell after the five-year usage rule. Whether they are showing wear or not, rotate them out to prevent odor buildup. Wearing improper shoes can result in fluid issues in and around the brain.

Footwear is to be worn in all toilet areas, including swimming pool locker rooms. When your day comes to a close, be sure to wash any excess soil or mud off shoes before storing them for the night.

House slippers with no coverage around the heel, flip-flops, and dress shoes with no coverage around the heel all create genetics for large feet. Any form of open shoe or slip on shoe is a message that speaks of instability and the walk will eventually become wobbly with a sense of insecurity. Have anyone in the family who seems to trip and fall easily? This may be a reflection of great-grandma's selection of slip-on shoes. House slippers should be made of sheep's wool, or other natural fabric, no plastic soles, fleece or terrycloth interior.

Shoes or accessories made from reptile hide carry the essence of that reptile. Not being a fan of the reptile family, not wearing or carrying articles made of reptile hide poses no hard feelings for me. Giving this some thought, it would make sense that this reptilian energy element hangs on to the hide. Something took place during the manifestation of a reptile (or anything else) creating their shiny, slick, scale-ridden baldness they display. I think I'll pass on the chance of being the recipient of signals that exist within the reptile (bald, scaley) category.

Back to the 1950's look: White socks are preferred, and ivory socks are permissible in the correct context. Socks should be worn with shoes, especially when going out of the home. Winter socks can ride above the knee and be made of cotton, silk, wool, cashmere or alpaca. Summer socks are to fold at the ankle and be made of cotton. Socks are worn to protect the meridians in the feet and should be clean each day. Wearing dirty socks opens the door to genetics for homosexuality and, red hair. Avoid the elastic socks claiming to improve circulation and no tube socks (those that come to the knee). It is best for men's dress socks are to rest at the top of the ankle, not a crew or knee sock.

The body does not seem to like shoes laces. Shoelaces can easily become untied resulting in a tripping hazard. Buckles on shoes provide a message to the body of security and stability.

The Small Stuff

The Victorian Era started the ball rolling for the faux feather boas that were a popular accent to the lady's wardrobe. Were faux feather boas just for the "ladies of the evening"? The Edwardian Era, 1920 flapper girls and the 1970s and 1990s contributed their share in the popular displays of the boa feathers as well. These fluffy tails of frizz that little girls would play dress-up with contribute to genetics that result in large breasts with no defining separation from the boobs to the belly! Great! Thanks Queen Victoria. This scenario might cause you to question whether your great grandmother was a show girl! What a thought! Grandma thought her days of being the Lady of the Evening would be kept secret. Everything eventually comes out in the open, albeit in a different form or fashion than what one might imagine. Ladies, it is not advisable to use boa feathers around the neck, or anywhere for that matter. Some items from the Victorian Era need to be eliminated from the wardrobes of today. We can start with elimination of the boa, then add the traditional white wedding dress, and throw in the dresses that sweep the ground. All need a serious makeover.

Luke 8:17: *For nothing is concealed that won't be revealed, and nothing hidden that won't be made known and come to light.* (HCS)

As minor as it may seem, buttons, embroidery and sequence all join in the fun of making or breaking a harmonious set of internal events. Shell buttons, those Mother of Pearl, shiny and pretty buttons send a water signal to the electrical system in the body. This can be troublesome when the buttons run along the spine, front or back. Pretty as they are, it is best to steer clear of the shell buttons.

Embroidery speaks of wealth. Back in the day when clothing spoke of status, it was considered an item for the wealthy to have embroidery on clothing or linens. Now days, sewing machines make it quite simple to add a touch of embroidery to an article of clothing. Embroidery does add a nice upscale touch when done in an elegant manner or design.

It cannot be stated enough, so repeated information is in order here. Lace and lace fabrics conflict with the function of the meridians leaving the door open for unwanted health interruptions. Lace leaves the spine exposed to invaders that pass through the body and will result in spinal fluid accumulation or even infection if given the opportunity. Black lace can be put in the dangerous category! Black has a powerful impact all its own and combining it with lace is a double-layered problem. Those lacey, frilly and fancy undergarments often seen on display at the local mall can initiate the development of bowel issues. It would be safe to say everyone would want to avoid such an issue.

Red silky bras and under panties will influence the pancreas and contribute to pancreatic cancers, fatigue and breast cancers. As you will recall, the color red targets its influence toward the blood.

Special occasions may warrant the use of nail polish. Nail polish has an impact on the physical body and wearing polish on the fingernails or toenails for extended periods of time could result in unwanted symptoms. Not only are nail polishes and polish remover toxic, but an element of color is also involved. Your hands touch numerous items during your day and transferring a signal from a toxin or color could build into unwanted health problems. Think of the preparation of food by hands with Acrylic nails. What could that be doing to the food? There are many things along these lines to be considered. Yes, manicure your nails appropriately. When nails are properly cared for and you are healthy, there is no need for nail polish because the finger below the nail will have a natural pink hue. Even the small stuff counts.

John 15:18-19: *"If the world hates you, understand that it hated Me before it hated you. If you were of the world, the world would love you as its own. However, because you are not of the world, but I have chosen you out of it, the world hates you."* (HCS)

CHAPTER XVIII

FROM BEACHES TO SKI SLOPES

Electricity to the heart is influenced by bikini or other two-piece style bathing suits. This would include the strapless tops on bathing suits. The shoulders do not like being exposed, especially to the outdoors. Shoulders are symbolic for carrying a burden. Bikini style suits cause a genetic imprint that increases the bustline.

Remember the discussion about exposing the legs with a short skirt or a pair of shorts? Wearing a bathing suit qualifies for the same list of consequences, genetics for fat or thick thighs, and hair growth on the legs.

Thong style swimsuits or underwear will generate genetics for a large derriere, whether for you or for future generations. Grandma thought you would never know about her skimpy bathing suit! There are no secrets in this game of life.

There are spiritual laws concerning exposing yourself to others and this includes the lack of modesty displayed at a public swimming pool or beach, not to mention nude beaches. People are harming themselves, future children or grandchildren when too much skin is exposed not only to others around them but to the elements in the atmosphere. Chemical element #32 Germanium comes to mind, and there could be many others that are influenced by the lack of coverage on the body. Too much skin exposure will cost a spiritual price.

Bundle Up

Keeping the body adequately warm can be a challenge at times, depending on the climate in which you live. Avoiding the popular synthetic jacket or coat, filled with Polyfil, duck or goose down or not, not only generates unwanted static electricity but blocks the body from receiving the chemical elements in the atmosphere. That puffy over-exaggerated look your coat displays may transfer and your body will look like your coat. Interior liners of jackets or coats are often made from a synthetic material as well. Do your best to avoid the slick Polyester or Rayon liners. Easy to clean and great for those who may be a little messy, or are they really that great? It is possible to have the liner inside a jacket or coat professionally replaced although it can become costly. Eventually the fashion industry will catch up to the idea of "no synthetics." Likely when people stop purchasing it. Coats should ride no lower than just above the knees. Wool, alpaca or cashmere gloves, scarves and hats are great for those outings in the cold.

Not-So-Fun Fact: Cape style coats with belts at the waist cause the "Superman" to die, meaning your ability to move in or with the Spirit Energy will die out.

CHAPTER XIX
THE VALUE OF BEING AUTHENTIC

Ladies, you are going to love this chapter! There is an influence from authentic gemstones and gold that has remained under the radar for several years, at least in the United States. Toss out the costume accessories because it is the real stuff that produces the look a lady will long for.

Authentic gemstones attract the chemical elements and result in an engagement with and development of the Spirit Energy within the body. This jewelry concept goes hand-in-hand with Matthew 11:8 in Chapter I, The Language of Clothing, mentioning how the way you dress is a reflection of your status, or in what manner you live your life. Jewels, like color, have seasons that will increase or decrease their influence. This I am still learning.

Color Selections

Revelation 21:19: *And the foundations of the wall of the city were garnished with all manner of precious stones. The first foundation was jasper; the second, sapphire; the third, a chalcedony; the fourth, an emerald; the fifth, sardonyx; the sixth, sardius; the seventh, chrysolite; the eighth, beryl; the ninth, a topaz; the tenth, a chrysoprase; the eleventh, a jacinth; the twelfth, an amethyst.* (HCS) (A wall represents a form of protection or barrier from something; an immoveable division.)

The color and vibrational quality of a stone will benefit the interior function of the body. The color of each stone is the result of a combination of the chemical elements from which it is made. Could wearing a particular color influence the body in a similar way as the same colored stone would have? For example, if you wear Emerald, will your body attract the chemical elements that form Emeralds?

Jasper (red, yellow, brown or green)
Sapphire (blue, white, pink, yellow, orange, green)
Agate/Chalcedony (Blue/grey quartz)
Emerald (bright green)
Sardonyx/onyx (black with white streaks)
Sarduis/Ruby (blood reds)
Chrysolite (milky light green)
Beryl (green)
Topaz (light purple)
Chrysoprase – contains nickel (milky mint green)
Jacinth (yellow, brown-red tones)
Amethyst (purple)

Exodus 28:31-34: *You are to make the robe of the ephod (ephod has a Hebrew origin meaning "to put on") entirely of blue yarn. There should be an opening at its top in the center of it. Around the opening there should be a woven collar with an opening like that of body armor so that it does not tear. Make pomegranates of blue, purple and scarlet yarn on its lower hem and all around it. Put gold bells between them all the way around, so that gold bells and pomegranates alternate around the lower hem of the robe.* (CSB) (description added)

These verses give a good example of how important clothing color, and mineral or precious stones can be. The ephod held precious stones that were worn over the region of the chest. Notice the reference to body armor? A form of protection is provided to the body when mineral or precious stones are implemented into the daily wardrobe. Mineral stones attract the necessary chemical elements from the atmosphere and produce the combination of gases the body can use. The Heavenly Gases help protect from influence originating from charged clouds (elemental forces) I spoke of in previous publications, that may be too strong for our body to safely encounter. We see a glimpse of the power of the precious mineral stones in Luke 19:40:

He answered, "I tell you, if they (people) *were to keep silent, the stones would cry out."* (CSB) (description added)

The physical body is to house a function that brings forth a sound that originates from vibration and that sound would reach the Heavens. Evidence of this comes through the reference of "cry out." The terms sing or dance found in Scripture are reference to an internal orchestration of events that produce a sound beyond what is normally heard by the human ear. Is there some form of a sound meter that could measure this music that is orchestrated within the body, or is it only for the spiritual ear to hear?

Union and Precious Stones

Jeremiah 2:32: *Can a young woman forget her jewelry or a bride her wedding sash? Yet My people have forgotten Me for countless days.* (HCS)

Scripture describes being joined to or having a union (dual agreement) with God by the term bride, groom, wedding, and so forth. As Jeremiah states, if a bride needs her jewelry, then this tells me that the jewelry is what attracts a visitation from the chemical elements (aka God) that creates the health protecting combination I call Heavenly Gases.

Add Revelation 22:17 below to this and we not only have a bride reference but a Living Water reference (self-hydration). I expand on the topic of Living Water in previous publications, but a simplified explanation is the proper measurement of Hydrogen + Oxygen inside the body will produce self-hydration. This verse is an indication that authentic jewelry, not costume jewelry, contributes to the production of the Living Water.

Revelation 22:17: *Both the Spirit* (Heavenly Gases) *and the bride say "Come!" Anyone who hears* (to understand) *should say "Come!" And the one who is thirsty should come. Whoever desires should take the living water as a gift.* (HCS) (description added)

Sparkling stones and gold trigger an activity within the body that will eventually produce certain youthful and attractive physical features. I can see the lines to enter the jewelry store growing already! As we unfold some of the relative Scripture, we may also get a glimpse of why the ephod was made for the priests, or should the story state how the ephod graduated someone into priesthood?

Proverbs 20:15: *There is gold and a multitude of jewels but knowledgeable lips are a rare treasure.* (HCS)

Proverbs hints that when jewels (some translations specify rubies), are set in gold, they produce "knowledgeable lips." Knowledgeable lips would be a product of a healthy functioning hippocampus, the learning center of the brain.

Sprinkled throughout the book of Song of Solomon are a few jewelry related Scriptures:

Song of Solomon 1:10: *Your <u>cheeks</u> are beautiful with jewelry (earrings), your <u>neck</u> with its necklace. We will make gold jewelry for you, accented with silver. 4:9: You have captured my heart, my sister, my bride you have captured my <u>heart</u> with one glance of your eyes, with one jewel of your necklace. 5:12: His <u>eyes</u> are like doves beside streams of water, washed in milk and set like jewels. 5:14: His <u>arms</u> are rods of gold set with topaz. His body is an ivory panel covered with sapphires. 7:1: How beautiful are your sandaled <u>feet</u> princess! The curves of your <u>thighs</u> are like jewelry, the handiwork of a master.* (CSB) (emphasis added)

I have placed an underscore on the words that need attention. There is a pattern of jewels or gold and silver being connected to a specific physical feature, and even the heart. Cheeks are to be full and the neckline wrinkle-free. Bracelets that rest on the left wrist will fuel the meridians that lead to the heart so the statement of the heart being influenced is no surprise. These few verses reveal much about the energy influence present in gold, silver and precious stones (authentic jewels). What could all this tell us about the overall health of the body?

Proverbs 31:10: *Who can find a wife of noble character? She is far more precious than jewels.* (CSB)

It is important to remember: Proper proportions of various chemical elements will produce light within the cells. The light eliminates infections and inflammation, and roots of many other health related issues.

Let's look at some of the individual precious stones.

Rubies

Isaiah 54:12: *I will make your <u>fortifications</u> out of <u>rubies</u> your <u>gates</u> out of <u>sparkling stones</u> and all your <u>walls</u> out of <u>precious stones</u>.* 61:10: *I greatly rejoice in the Lord, I exult in my God; for he has <u>clothed me in salvation</u> and <u>wrapped me in a robe of righteousness</u> as a groom wears a <u>turban</u> and as a bride adorns herself with <u>her jewels</u>.* (CSB) (emphasis added)

Again, focus on the underscoring. Fortifications represent strength. Rubies provide a level of strength for the body. Gate(s), a term to describe an action, process or mechanism by which the passage of something is controlled; a point of entry or initiation of something. It could be said that the verse is relative to blood pressure, a measurement of the rate of rhythm and many other fluid related activities that take place inside the body or, the opening and closing of the cells. Salvation is a means of rescue. The jewels are providing not only a steady operation of the functions inside the body but also a rescue from unforeseen interruptions in health. I'm going to interject my groom and bride descriptions here. Groom is a reference to the physical functions or material, physical body; that which comes from the soil/earth. Bride is the Spirit activity (Heavenly Gases), not necessarily seen but joins to the physical body causing the physical and spiritual to join together. Simple physics. The "groom wears a turban" is a reference to the brain, turbans cover the head, and the description of righteousness presented just prior to the word "groom" tosses in a hint of the decision-making processes being involved. Men receive decision-making guidance through gemstones while women receive beauty and fuel for Spirit enlightenment. The short version is jewelry is taking care of many physical processes on the interior of the body currently unrecognized by man.

Diamonds

Ezekiel 3:8-9: *Look, I have made your face as hard as their faces and your forehead as hard as their foreheads. I have made your forehead like a diamond, harder than flint. Don't be afraid of them or discouraged by the look on their faces even though they are a rebellious house.* (CSB)

Ezekiel tells us that Diamonds are a protector of the frontal lobe part of the brain. There will be times when people oppose you or your situation but your frontal lobe functions kick in and what God has taught you will be a weapon of the warfare at hand.

Sapphires

Ezekiel 28:13: *You were in Eden, the garden of God. Every kind of precious stone covered you: carnelian, topaz and diamond, beryl, onyx and jasper, sapphire, turquoise and emerald. Your mountains and settings were crafted in gold; they were prepared on the day you were created.* (CSB)

Ezekiel 10:1: *Then I looked, and there above the expanse over the heads of the cherubim was something like sapphire stone resembling the shape of a throne that appeared above them.* (CSB)

Lamentations 4:7: *Her dignitaries were brighter than snow; whiter than milk; their bodies were more ruddy than coral (rubies), their appearance like sapphire.* (CSB)
Isaiah 54:11: *Poor Jerusalem, storm-tossed, and not*

comforted, I will set your stones in black mortar and lay your foundations in sapphires. (CSB)

Ezekiel gives a fair amount of information and gives rise to the notion that if people desire to return to an "Eden" style of living, the precious stones will have to be implemented into daily life. Taking this a step further, the extreme weather changes experienced around the globe may begin to experience a downsize in their intensity simply through mankind receiving the necessary elements from authentic jewels, processing what they receive through an act of photosynthesis, and giving back to the earth. Eden is not described as having drastic weather or temperature changes. Only reference I have come across is that the "cool of the day" was evening, once the sun went down. Wouldn't it be simply glorious to have a more regulated weather pattern around the globe! This would be equivalent to Paradise.

In Lamentations, snow and milk reflect fair, smooth, clear skin introduced by sapphire stones. Rubies, or in some translations it states coral, is a reference to a tone body with nothing sagging, bagging or flabby. Simply said, sapphires and rubies help keep the body tone. Bye, bye gym memberships!

Isaiah mentions foundations, a reference to the physical characteristics of a person. The foundational things such as shape and size the physical body has. Sapphires provide the stability as seen in the reference to being set in black mortar.

Let's expand the idea that our internal spark came forth by the works of the precious stone. Is human life a result of the synergetic influence precious stones emit into the atmosphere? That twinkle or sparkle is what I call

Star Dust that resides in the plasma and becomes more evident when contamination is removed from the plasma. Not a question I can answer but possibly one someone will address at a time in the future.

Jeremiah 1:5: *Before I formed thee in the belly I knew thee; and before thou camest forth out of the womb I sanctified thee, and I ordained thee a prophet unto the nations.* (KJV)

Beryl and Topaz

Ezekiel 1:16: *The appearance of the wheels and their craftsmanship was like the gleam of beryl and all four had the same form. Their appearance and craftsmanship was like a wheel within a wheel.* (HCS)

Daniel 10:6: *His body was like Topaz, his face like the brilliance of lightening, his eyes like flaming torches, his arms and feet like the gleam of polished bronze and the sound of his words like the sound of a multitude.* (HCS)

Ezekiel is speaking of the chakras (wheels) and when influenced by the precious stone beryl, function properly.

Daniel describes the physical changes that come with exposure to Topaz. The face will shine or glow, the eyes will be blue and the arms and feet will be strong, stable providing a protection during the walk of life.

Pearls

Pearls equate to wisdom. Pearls are not listed as a precious stone as found in Ezekiel 28:13. Pearls are not a product of the chemical elements within the soil and due to their roots of formation, are an influence to the fluid activity in the body, helping the cleansing activity through the lymph system. There are specific months Pearls should be worn with focus being on a benefit in March/April dependent upon the designated date for Easter. When additional information about jewels and seasons or months of their benefit are better had, I will pass that information along.

Precious Stones

When cells are properly fueled by the chemical elements, disease is challenged at taking over the cell. I'm not going to say no one will ever encounter a health concern, there's simply too many factors and variations that can play into an individual situation. Precious stones aid in providing the combination of chemical elements the cells need to stay clean, well-functioning and disease free.

Exodus 24:10: ...*and they saw the God of Israel. Beneath his feet was something like a pavement made of lapis lazuli* (deep blue metamorphic semi-precious stone), *as clear as the sky itself.* (CSB) (description added).

••   ••

Rings

Rings should fit the finger and avoid excess movement, twisting or turning. This rotation movement of a ring and rings that are designed in a spiral with no connecting ends plant seeds for rheumatoid arthritis, a crippling of the hands. Playing a piano will also result in the development of genetics for rheumatoid arthritis.

Watches

Wristwatch versus pocket watch. Heart and lung meridians run through the arms down to the tips of some of the fingers. Wearing a watch on the wrist will interfere with the function of these meridians and through time and generations will develop some unwanted electrical issues in the forebrain resulting in the loss of intelligence. It would be wise to accept and put back into action the old pocket watch. The more you can avoid handling the cell phone and wearing the Apple wristwatch the better your energy throughout the day will be. You might be a whole lot smarter for it as well.

Not-So-Fun Fact: A lack of wearing authentic jewelry results in leaking gut.

CONCLUSION

<u>IN THE CLOSET</u>

Your closet may hold numerous choices
Sometimes it proves best to stifle fashioner's voices

All settled into their proper places
For your review at the dawn of days paces

Some nicely folded and some hung for display
The color I choose could make this a great day

Your power is increased or lost quite quickly
By the choices you make so it's best to be picky

The combination you select will give you a fix
From basics to details it's all in the mix

Stripes of red, with white and with blue
Who would have known it produces the flu

When the bum begins to droop, and the legs become
flabby
Aging crept up, a good reason to be crabby

Poor clothing choices dominated the days
Of Grandma and Grandpa and things went astray

The body remains fit with the proper attire
No need to run or jump to create a perspire

To achieve a measure of timeless aging
Follow these steps even if they sound crazy

Give it some time and you will see
The body become fit and wrinkle free

Colorful dresses and two-inch heels
Hold a few secrets to how one will feel

Jewelry, bows and colorful sashes
To aid the body and prevent energy crashes

Belts, shoes, ties and more
What should I select to produce even more

Wear a hat to shade from the sun
The crown needs protection to properly run

Cover the feet with socks and with shoes
Expose the toes and the power you lose

Hold fast to the secrets of what this will do
For no one else will have even a clue

Color and style are what I support
This all contributes to my now good report!

♛

Izauh 61®

*We have no way of being ourselves
when the construct of our being is comprised
of the lifestyle choices of our ancestors.*

RESOURCES

1)   Holy Bible: Holman Christian Standard; Christian Standard Bible; New King James Version
2)   *The Wisdom of Solomon*, NR Publishers 2024
3)   *The Books of Enoch*, Published 2024
4)   Publication@medium.com

Suggested Reading:

From AntiChrist to I AM
Food for the Journey to I AM
       Published 2022, Harvest of Healing, LLC

Home-Made Answers for Cancer and Life Altering Disease
       Published 2024, Harvest of Healing, LLC
Living by the Light of the Moon Published 2024, Harvest
       of Healing, LLC
Eating Yourself to Death, Published 2024, Harvest
       of Healing, LLC

www.ingramcontent.com/pod-product-compliance
Lightning Source LLC
Chambersburg PA
CBHW031548260326
41914CB00002B/326